W9-BKJ-657

BODYRHYTHMS

BODYRHYTHMS

CHRONOBIOLOGY AND PEAK PERFORMANCE

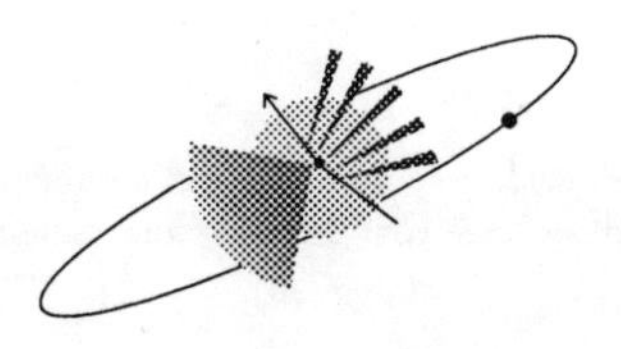

LYNNE LAMBERG

WILLIAM MORROW AND COMPANY, INC.
New York

For Stanford

and

Nicole and Ryan

Although every attempt has been made to ensure that the information in this book is accurate, it does not constitute medical advice. See your doctor before using medication or other treatment.

It is the policy of William Morrow and Company, Inc., and its imprints and affiliates, recognizing the importance of preserving what has been written, to print the books we publish on acid-free paper, and we exert our best efforts to that end.

Library of Congress Cataloging-in-Publication Data

Lamberg, Lynne.
Bodyrhythms : chronobiology and peak performance / Lynne Lamberg.
p. cm.
Includes bibliographical references and index.
ISBN 0-87795-991-9
1. Biological rhythms. I. Title.
QP84.6.L35 1994
612'.022—dc20 93-40320
CIP

Printed in the United States of America

First Edition

1 2 3 4 5 6 7 8 9 10

BOOK DESIGN BY MICHAEL MENDELSOHN/MM DESIGN 2000, INC.

FOREWORD

THIS IS THE BEST BOOK that has ever been written on the human journey through the nychthemeron. Nychthemeron is my favorite obscure, yet very useful word. It means the entire twenty-four-hour day: the day and the night in their most intimate functional relationship. In other words, this is the most comprehensive text on sleep and wakefulness that has ever been written for the lay reader, and Lynne Lamberg, its author, is as much of a scientific expert as any nonscientist could be.

In 1984, her book, *The Guide to Better Sleep*, was published by the American Medical Association as part of its Home Health Library Series. It was the state of the art publication for the lay public. I was one of the advisers on that book. I worked with Lynne Lamberg extensively and I have continued to do so ever since. She has attended most scientific meetings in the field, visited many sleep laboratories, and written countless articles on sleep. She is, in sum, superbly qualified.

Although the discipline of sleep disorders medicine or somnology is now well established in its scientific knowledge base and clinical practice, it nonetheless is in the backwaters of neglect in terms of a mandate derived from pervasive societal understanding. Unlike research in many other areas of medicine, such as cancer, where most people have a reasonable understanding of the illness and its danger signs, and of the value of research, few people have a correct understanding of the principles of sleep and wakefulness.

Human beings have always been aware of the differences between day and night, but it is only relatively recently that scientists have realized that timekeeping is internalized. I would venture to say that biologic clocks are in the public vocabulary now. But the precise sleep-related determinants of daytime alertness have only lately been clarified, and they are not yet in the public lexicon.

This is the first book for the lay public that adequately integrates the two major areas of sleep/wake regulation: the circadian timekeeping system, and the homeostatic regulation of sleep and wakefulness, which works very much like thirst and hunger, with deprivation intensifying consummatory behavior. There is no other introductory text that describes or deals with this important dimension of cognitive function and human performance.

As Lynne Lamberg documents graphically and convincingly, impaired alertness is one of the greatest potential dangers of contemporary life. It is about time that a book delves into the interconnectedness of sleepiness/alertness and its profound effects on our lives and our civilization. This book, we hope, will educate the American public on this vital subject.

In terms of public awareness, in the area of personal health, there are three key domains: physical fitness, nutrition, and, finally, good sleep and alertness. There have been hundreds if not thousands of books on physical fitness and nutrition. This book is the first that does justice to the third pillar of this crucial triumvirate.

—William C. Dement, M.D., Ph.D.

Director, Sleep Disorders Clinic and Research Center
Lowell W. and Josephine Q. Berry Professor of
Psychiatry and Behavioral Sciences
Stanford University School of Medicine

Chair, National Commission on Sleep Disorders Research

ACKNOWLEDGMENTS

MANY SPECIALISTS IN BIOLOGIC RHYTHMS and sleep research were wonderfully generous with their time, offering access to their laboratories and details of studies still in progress. Most even returned phone calls promptly! Some reviewed parts of this book and offered valuable suggestions for changes and additions. William Dement, M.D., Ph.D., Stanford University, inspired me with his passionate dedication to both the study and treatment of disorders of biologic rhythms and sleep and to the dissemination of the latest findings to the public. Other scientists to whom I am enormously indebted include Roger Broughton, M.D., Ottawa General Hospital, Ottawa, Ontario, Canada; Mary Carskadon, Ph.D., E. P. Bradley Hospital, East Providence, Rhode Island; Michael Chase, Ph.D., University of California at Los Angeles; David Dinges, Ph.D., Institute of Pennsylvania Hospital, Philadelphia, Pennsylvania; Charmane Eastman, Ph.D., Rush-Presbyterian-St. Luke's Medical Center, Chicago, Illinois; R. Curtis Graeber, Ph.D., Boeing Commercial Airplane Group, Seattle, Washington; Franz Halberg, M.D., University of Minnesota, Minneapolis, Minnesota; Daniel F. Kripke, M.D., San Diego Veterans Administration Medical Center, San Diego, California; Kathryn Lee, Ph.D., University of California at San Francisco; Alfred Lewy, M.D., Oregon Health Sciences University, Portland, Oregon; Margaret Moline, Ph.D., New York Hospital-Cornell Medical Center, White Plains, New York; Timothy Monk, Ph.D., University of Pittsburgh Western Psychiatric Institute and Clinic, Pittsburgh, Pennsylvania; Charles Pollak, M.D., New York Hospital-Cornell Medical Center, White Plains, New York; Norman Rosenthal, M.D., National Institute of Mental Health, Bethesda, Maryland; John C. Urbaitis, M.D., Sinai Hospital, Baltimore, Maryland; and James Walsh, Ph.D., St. Luke's Hospital, St. Louis, Missouri.

Rose Marie Morse of William Morrow became the editor of this book midway through. She took on the task with sensitivity and zeal, providing thoughtful direction for revisions and shepherding the book through the production process. Thank you, Rose Marie. My thanks also to copyeditor Sonia Greenbaum, whose meticulous attention to detail enriched the readability of virtually every page. I am grateful to my agent, Sandra Dijkstra, for her confidence and seemingly inexhaustible energy.

Additionally, I am grateful to Muriel Weitzman for an insider's guide to who's who in chronobiology, for bed and breakfast on my numerous visits, and for her gift of books from the library of the late Elliot Weitzman, her husband, a pioneer in chronobiology and one of my mentors in the field. Toba Cohen, Jane Alexander, and Elisa Petrini gave me constructive advice on organization and style.

Many friends provided anecdotes, newspaper clippings, and attentive ears. They include Walter Bonime, Doris Diamond, Ricky and Eric Fine, Carol and Sandy Frank, Jill and John Lion, Nancy Nyman, and Carole Simon. Sandy Hillman tested the jet lag coping guide. My fellow writers and members of The Midlist Club are eager for me to pick up the dinner check.

Thanks also to the astronomers, doctors, nurses, sleep laboratory technicians, flight attendants, firefighters, police officers, broadcast reporters, suicide hotline volunteers, actors, power company employees, and others who vastly enlarged my understanding of what it means to work at odd hours, particularly at night. I hope to include their individual stories in another book.

Special thanks to the many librarians who aided my research, particularly those at the telephone reference service and other departments of Baltimore's Enoch Pratt Free Library who kept going on many trails I had supposed were cold. " 'Impossible' is not in my vocabulary," one told me.

Finally, my gratitude to my husband Stanford, who listened to tales of my adventures, alerted me to articles in medical journals, served as an astute first reader, and even consented to sleep with the shades open so that I could set my biologic clock with morning light. Thanks also to my children, Nicole and Ryan, who good-naturedly tolerated my advice about their sleep habits even if they seldom followed it.

CONTENTS

PART I

HOW BODY CLOCKS WORK

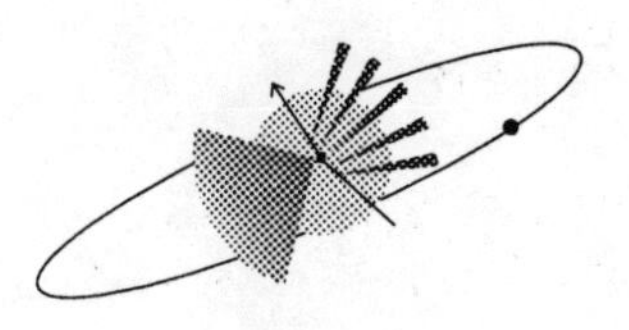

CHAPTER 1

THE BIOLOGIC CLOCKSHOP

TO ENGAGE IN NIGHT WORK is to strike a Faustian bargain. A price must be paid for the use of the night and the day in ways for which the human body was not designed. People who stay awake at night, when the body calls for sleep, are less alert than they would be during the day, and are more likely to make mistakes. People who sleep during the day, when the body churns with wakefulness, sleep less soundly than they would at night, and they do not sleep as long. Chronic sleep deprivation in and of itself will impair performance, day and night.

Rotating between day and night brings added tribulation: changes in hours of work usually necessitate changes in hours for sleeping, and in other daily activities. Workers may be ready for breakfast, for instance, when their families are sitting down to dinner. People who work at night, or change their schedules frequently, have higher incidences of sleep disorders, digestive disorders, mood disturbances, heart disease, and other health problems than their dayworking peers; their family life and social life often suffer as well.

Yet one in five workers in the United States—and an estimated one in seven in industrialized nations worldwide—is on the job at hours other than nine to five.

Butchers, bakers, and automakers, even doctors, lawyers, and corporate chiefs, all may work around the clock. Until recently, the night shift was populated almost exclusively by blue-collar workers, but modern technology and global communications have brought more and more professionals into the twenty-four-hour workforce. This is a change with major social and political ramifications, for it has raised public awareness of shift-work issues, moving them from the back burner of the nation's legislative stovetop to the front burner.

The new American shift worker may practice shuttle diplomacy, deal on the international monetary market, or report breaking news. He or she may hold direct responsibility for hundreds of lives, and for costly equipment, piloting large jetliners, for example, or monitoring the control panels at nuclear power plants.

Today, people at all levels of society, and in all types of occupations, can expect to spend at least part of their lives working on the night shift. In the last twenty-five years, the around-the-clock workforce in the United States has doubled, and the rate of increase is accelerating rapidly. When television stations, brokerage houses, and bookstores operate twenty-four hours a day, so will child care centers, banks, hairdressers, and even dentists' offices.

Society pays a substantial price for pushing the human machine beyond its design limitations: major industrial catastrophes have occurred more often at night than during the daylight hours.

The *Exxon Valdez*, for example, ran aground at 12:04 A.M. on March 24, 1989, in well-charted waters off the coast of Alaska, with no fog and no other traffic nearby. It spilled eleven million gallons of crude oil, despoiling wildlife, the area's fishing industry, and twelve hundred miles of pristine shoreline.

The news media gave a lot of play to the captain's absence from the bridge at the time of the accident, and to his impairment from alcohol. They also highlighted the inexperience of both the third mate the captain had left in charge, and of the seaman at the wheel. Few news reports, however, noted that the crew members' usual work schedule—six hours on duty alternating with six hours off duty—kept them from ever getting more than five hours of unbroken sleep. Moreover, they often worked virtually around the clock, thus compounding their already stressed state. On the night before the accident, the National Transportation Safety Board (NTSB) concluded, the third mate probably got about four hours' sleep. He then worked a physically arduous day, with a brief nap between his afternoon shift and supper. Thus, in the twenty-four hours before the accident, the third mate could have had less than five hours of sleep.

In its report on the *Valdez* accident, the Coast Guard said sixteen-hour workdays were common in shipping operations. It is not unusual for some mates to be so exhausted when they arrive in port that they cannot even stay awake long enough to talk with Coast Guard officers to set up the required inspections.

Perhaps the most devastating effect of sleep loss is its subtle and insidious effect on vigilance. Sleepy people miss things they would seldom overlook when well rested. The *Valdez* accident was one of omission, not commission. While crew members did not drive the ship onto the reef, they did however not respond to warning lights. According to the NTSB, "the third mate's failure to turn the vessel at the proper time and with sufficient rudder probably was the result of his fatigued condition. . . ."

Such problems are not limited to shipping. All major calamities in the nuclear power industry have occurred during the sleepiest hours of the day. The biggest nuclear accident in the United States began between 4 and 6 A.M. on March 28, 1979, when workers in the nuclear power plant at Three Mile Island in Pennsylvania failed to notice that a valve was stuck. This lapse triggered a major loss of core coolant water and the near meltdown of the reactor, which workers only narrowly averted. Significantly, just that day the workers had rotated from the day shift to the night shift. "Except for human failures," says the report of the President's Commission on the catastrophe, "the major accident at Three Mile Island would have been a minor incident."

At 1:23 A.M. on April 26, 1986, the reactor at the Chernobyl nuclear plant, near Kiev in the Soviet Union, spewed high levels of radiation in history's worst nuclear power accident. At least three hundred people are known to have died. Radioactive clouds spread contamination to millions more, both nearby and in surrounding European countries. Four years later, residents of contaminated towns had higher than normal complaints of fatigue, loss of vision, and loss of appetite, all symptoms of radiation poisoning. Farmers in the area reported births of horses with eight legs, and of pigs with no eyes. Soviet officials blamed the accident on faulty equipment design and on "awkward and silly actions by operators." Engineers involved in the disaster had been at work for thirteen hours or more.

Work schedules also played a role in the 1984 chemical disaster at a Union Carbide plant in Bhopal, India. This accident occurred at 12:40 A.M. on December 4, 1984, just after a shift change. Workers on the new shift failed to notice that pressure in a pesticide chemical tank had increased fivefold between the last reading of the previous shift, and the first reading of their shift. Water, leaking into the tank through a faulty valve, mixed with the chemical, producing pressure sufficient to rupture the tank and release poisonous gas. A low cloud of the gas drifted over the nearby city, killing three thousand people and injuring thirty thousand, according to Indian government reports.

In an increasingly twenty-four-hour world, many people not only expect to be able to sleep whenever convenient, but they also view sleep as expendable, an activity that can be dispensed with if more interesting, or more urgent, business is at hand. Such beliefs have been a frequent ingredient in major disasters.

The space shuttle *Challenger* exploded just after its noontime liftoff on January 28, 1986. It was witnessed by millions of horrified Americans, many of them children. The presidential commission that investigated the disaster focused on the failure of rubber O-rings on the booster rocket to form a tight seal, permitting flames to escape and set off the explosion that killed all seven aboard, including a teacher, Christa McAuliffe, who was to have taught the nation's schoolchildren lessons from space. At the hearings, physicist Richard Feynman dunked a piece of a rubber O-ring in a glass of ice water to demonstrate its lack of resiliency, a simple experiment that made the impact of low temperature bone-chillingly clear.

When temperatures at the launch site plunged below freezing, there was "a clear opportunity for postponement," the commission concluded. "The decision should have been based on engineering judgements," it noted. "However, other factors may have impeded or prevented effective communication and exchange of information." The report cited the launch crew's excessive overtime, irregular working hours, insufficient sleep, and fatigue as possible contributory factors. Some key managers had had less than two hours' sleep the night before, and had been on duty since 1 A.M.

It was only after a previous shuttle launch had been scrubbed on January 6, 1986, that anyone realized that eighteen thousand pounds of liquid oxygen fuel had been inadvertently dumped. Weary console operators in the launch control center had misinterpreted computer warnings. These workers had been on duty for eleven hours for the third night in a row of a twelve-hour shift.

The economic impact of these and other time-of-day–related human errors is staggering. The cost of the cleanup of the *Exxon Valdez* oil spill alone has exceeded $3 billion so far, with billions more pending in lawsuits. Union Carbide agreed to provide $465 million to victims of the Bhopal accident.

"The failure to cope with human factors may be the Achilles' heel of modern industrial and military technology," Charles Czeisler of Harvard University warned a congressional subcommittee chaired by then Representative, now Vice President Albert Gore, Jr., during its investigation of shift work and biologic clocks in 1983. While technology leaps rapidly into the future, Czeisler said, "there has been relatively little progress in overcoming the human limitations of operating machines twenty-four hours a day."

The catastrophes cited above prompted extensive news media coverage and much hand-wringing, along with the predictable calls for more scientific study to investigate what went wrong. But little substantive action was taken to prevent future disasters. The long-term consequences of work schedules that flout the body's natural rhythms could become the twenty-first century's most significant public health problem, claiming more lives, and exacting a greater economic toll, than AIDS, cancer, and all other diseases combined.

There is hope of countering this forecast, however, and it lies in the young science of chronobiology, the study of internal timing mechanisms that govern not only sleepiness and alertness, but also all other aspects of daily life, from birth to death, in sickness and health, on the job and off it.

A NICHE IN TIME

One may view earth as a giant condo with time-shares, where all plants and animals have their designated periods of occupancy. Some species, including humans, have been assigned to the daytime, and others, to the night. This assignment, while not immutable, is not very malleable. Psychologist Wilse Webb of the University of Florida spent five years trying to teach rats not to sleep between noon and 6 P.M. "They, by their contrary nature," he recalled, "spent five years teaching me otherwise." Humans are able to function around the clock, but they remain diurnal creatures—a fact that must be reckoned with as society moves increasingly into a twenty-four-hour world.

All species were designed to be most energetic when their food sources are most accessible. According to nature's plan, honeybee activity coincides with the time of blossoming; and the worm's fateful encounter with the early bird is preordained. Internal clocks prompt bees to leave their hives, and birds their nests, so that they both arrive on time for their meals. They also ensure that the flowers open and the worms surface at the appropriate moment.

There is, indeed, a time for every purpose. As a legacy of evolution, all living organisms, from one-celled algae to human beings, are stamped with an inner world that mirrors the outside one. Plants and animals anticipate predictable events—sunrise, sunset, mealtimes—with the aid of automatic timers inside the body. These work much like timing devices in the home that adjust the heat or turn on the coffeepot at preset hours.

Before one awakens from sleep, for example, body temperature and blood pressure start to rise, and the secretion of "get up and go" hormones needed for waking activities, such as cortisol and adrenaline, begins. The ability to concentrate, to move quickly, and to use one's hands with dexterity, the desire to eat, and even the need to urinate peak during the usual waking hours; these functions plummet to their daily lows during the customary hours of sleep.

Hundreds, perhaps thousands, of cycles oscillate each day in all cells, tissues, and organs of the human body. Some peak early in the day, some in the middle of the day, and some late in the day. Nonetheless, these cycles are all synchronized with each other. The range between their highs and lows often is so sizable that it has been compared to being a blonde in the morning, a redhead at noon, and a brunette in the evening. Lewis Carroll's Alice could have been describing these cycles when she said "I know who I was when I got up this morning, but I think I must have been changed several times since then."

Just as cognitive performance is at its highest during the day, and lowest at night, the ups and downs of other daily cycles may affect health and well-being. Blood pressure, for example, may be 30 percent lower at 9 A.M. than at 6 P.M. Because of this difference, a person with high blood pressure may receive too little medication if he or she habitually stops by the doctor's office early in the day, while the same person, regularly visiting the doctor late in the day, may receive an inappropriately high dose. Similarly, most asthma attacks occur in the early morning, when one's airways are normally most constricted, and attacks are less frequent in the midafternoon, when airways are normally most relaxed and open.

THE DISCOVERY OF INNER CLOCKS

Some 2,300 years ago, Androsthenes, a scribe who accompanied Alexander the Great to India, noted that the tamarind tree opened its leaves during the day and closed them at night. Androsthenes concluded that the plant was worshiping the sun. This charming notion, although false, persisted for centuries. Until relatively recently, it was commonly thought that sunlight, and related daytime stimuli such as noise and heat, caused animals to awaken, while the lessening of these stimuli at night produced sleep.

The errors of this assumption became obvious only when the French astronomer Jean-Jacques d'Ortous de Mairan performed a simple but elegant experiment in 1729. He left a type of mimosa known as "the sensitive plant" in a dark closet. After seeing that it continued to raise and lower its leaves once a day, he concluded that leaf behavior originated within the plant.

But de Mairan was so busy with astronomical research that he did not take the time to record his seminal discovery, leaving it to a friend to present it at a meeting of the Royal Academy of Sciences in Paris. His peers did not appreciate the significance of his finding, now often cited as the dawn of the field of chronobiology. In the next two hundred years, it attracted interest mainly from botanists.

The Swedish naturalist Carolus Linnaeus said that a person could tell what time it was by walking around the countryside and noting which flowers were open or closed. In 1754, he designed a garden clock that told the time from 6 A.M. to 6 P.M. via beds of plants whose flowers opened at different hours by virtue of their own internal clocks. Flower clocks became a popular addition to many formal gardens. Some plants' names reveal the time of day they bloom: morning glory, four o'clock, and evening primrose.

The event that ushered in the modern era of chronobiology may have been the arrival in 1919 of a cage of rats at the Johns Hopkins University School of Medicine. The rats were waiting for Curt Richter, a young researcher, when he showed up for work one morning soon after coming to Hopkins to study with the famed behaviorist John B. Watson. Richter, a recent college graduate who had never seen rats up close, gave them milk and bread, and sat back to watch them.

"What impressed me most of all about the rats was their spontaneous activity: the fact that they just jumped all around the cage and climbed around for periods and then were quiet again," he recalled some years later. "I could not help but wonder what made them active.

"The source of this activity," he mused, "must lie somewhere within the organism." This important insight appeared in Richter's first paper, *A Behavioristic Study of the Activity of the Rat*, published in 1922. In hundreds of painstaking studies, he charted the natural patterns of sleeping, eating, drinking, exercising, and other behavior in rats and other mammals, bringing the concept of biologic clocks into public awareness.

Richter and other early researchers in chronobiology found that all the animals they studied in their laboratories, from sea slugs to monkeys, followed predictable daily rhythms. Regardless of whether the animals lived in constant light or constant darkness, with one meal a day or several, with changes in noise or activities, through all the manipulations the researchers could dream up, their daily rhythms persisted just as though they were living in their natural surroundings.

For many years, Richter staged a contest with the new students each July to see who could predict most accurately when a blind monkey living in the laboratory would sleep, and be active, on Christmas Day. Richter invariably won. He simply extended his chart of the animal's daily sleep/wake cycle, demonstrating to the students the high reliability of inner clocks.

Researchers concluded that animals did not learn their rhythms from observing others of their species. Lizards and chickens hatched in the laboratory, kept from contact with others, oblivious to the natural alternation of day and night, displayed the same daily rhythms as their parents. And so did their offspring. In one study, conducted by Colin Pittendrigh of Princeton University, the fifteenth successive generation of fruit flies, grown under a variety of artificial conditions, showed rhythms identical to those of the original generation. Their daily behavior patterns were inherited.

Researchers also concluded that daily rhythms were not a response to subtle environmental cues arising from the earth's rotation. They conducted experiments at the South Pole, where hamsters, fruit flies, bean plants, and fungi were kept on a turntable spinning in a direction opposite that of the earth. More recently, researchers even took fungi along on a space shuttle, keeping them away from external signals. Inner clocks kept on ticking as before.

DIFFERENT DRUMMERS

However, there was one amazing difference: life in the outside world followed a twenty-four-hour clock, conforming to the earth's rotation schedule. In the laboratory, away from light, dark, and other indicators of time, different species followed the dictates of different drummers. Most day-active animals and plants lived on days slightly longer than twenty-four hours. Owls, mice, and other night-active species lived on days that were somewhat shorter. Studies in humans, begun only in the 1960s, showed that most people have a natural internal rhythm of about twenty-five hours.

These daily rhythms are close enough to twenty-four hours to permit life to flourish in a twenty-four-hour world. Some plasticity is not surprising; the body has to be able to adapt to modest variations in its environment from one day to the next (on a rainy or overcast day, the early bird—or the worm—might be late) as well as to gradual changes in day length over the course of the year.

Across all species the natural day ranges from about twenty to twenty-eight hours; in most mammals, the span is between twenty-three and twenty-six hours. There is, of course, some variation among members of the same species, even between identical twins, even between conjoined twins. To describe daily rhythms, chronobiologist Franz Halberg of the University of Minnesota concocted the word "circadian" from the Latin words *circa*, meaning "about," and *dies*, "a day."

While the most conspicuous cycles in plants and animals are circadian, they intermesh with other cycles that are both shorter and longer than a day. There are rhythms within rhythms: the heart beats approximately once per second, but this rate speeds up in the morning, and slows at night. All living organisms are composites of rhythms with many frequencies, ranging from milliseconds to months, and even years. Among the shorter than one-day or *ultradian* cycles are the electrical activity of the human brain, which cycles in fractions of a second, and breathing, which occurs about sixteen times per minute. Many hormones are released in pulses lasting several hours.

Cycles with periods that are longer than one day are called *infradian*. Among these, weekly, or *circaseptan*, rhythms are the newest frontier in chronobiology research. Unlike daily, monthly, and yearly rhythms, the seven-day week has no obvious geophysical correlates. Many scholars have linked the week's origins to the biblical injunction to work for six days and rest on the seventh; they view the invention of the week as a social phenomenon.

It may have been the other way around, some chronobiologists now say. Astonishing as it might seem, what has long been thought a cultural convention may reflect the cadence of deeply buried biologic rhythms. Seven-day rhythms are prominent in many illnesses: the common cold, as one example, lasts about a week. Beard growth in men also follows an approximately weekly rhythm, as demonstrated by collecting and weighing the shavings from an electric razor; this observation, though seemingly trivial, may provide a window to further study of male hormonal rhythms.

The most familiar long infradian rhythms are the approximately monthly, or *circatrigintan*, rhythms, such as menstruation, and the approximately yearly, or *circannual*, rhythms, such as migration, hibernation, and reproduction. The former are tied to the lunar month, and the latter to the earth's rotation around the sun.

While annual rhythms are less prominent in humans than in other animals, they nonetheless exist. One might think that human sexual interest and activity depend too much on social interaction to show seasonal patterns, for instance. Not so. Spring, the poets say, is the season for lovers, but secretion of the male hormone testosterone surges in the fall, sperm concentration and activity is highest in winter, and rates of conception peak in winter, too. In the United States, more babies are born in August and September than in any other months—a pattern that government statisticians have documented with only a few exceptions in the last fifty years. Babies born in summer and fall tend to weigh slightly more than those born in other seasons, a rhythm that contributes to yet another: babies born in the summer and fall have a greater likelihood of survival.

Deaths, like births, have a seasonal pattern. Fatal heart attacks, for example, are more common in winter. It would be natural to suspect a link to cold-weather activities that strain the heart, such as shoveling snow or breathing frigid air, but the same winter death peak emerges in Hawaii and other places, where the temperature remains moderate year round, and even in the Southern Hemisphere, where seasons are reversed. Compared with the rest of the year, winter months are, however, a time of lower resistance to infections and of higher cholesterol levels, both of which may contribute to a higher risk of dying.

THE HOAX OF BIORHYTHMS

While it may seem logical to condense the term "biological rhythms" into biorhythms, specialists in chronobiology avoid the shortened form, because it has been coopted by those who subscribe to biorhythms theory. This pseudoscientific notion identifies purportedly critical days in a person's life solely on the basis of one's day of birth.

Because the theory offers a simple formula for gaining control over one's life, and offers the illusion of predictability, it has spawned a small industry devoted to providing biorhythms charts, books, and consulting services. There is one hitch: it does not work.

Despite occasional anecdotal reports of events that occurred as biorhythms theory had forecast—a football team, say, whose members scored more points on their so-called critical days—numerous scientific studies conducted in recent years showed that no relationship existed between highs and lows of the various biorhythm cycles and athletic performance; pilot errors; accidents in mines, factories, and other worksites; traffic fatalities, or admissions to hospitals.

Among long cycles in nature are those of the swallows that return to Capistrano, California, every March 19, including leap years (they have been late only twice in 200 years), periodical cicadas that live underground in mud cocoons while they wait for a wake-up call that may come as infrequently as once every 17 years, and a Chinese bamboo called Sleeping Beauty that blooms once every 125 years.

HOW THE MASTER CLOCK WORKS

In his attempts during the 1960s to find the source of control of daily rhythms, Richter removed minute slivers of tissue from hundreds of different areas of the rat brain. He found that injuring part of a tiny region at the center of the brain known as the hypothalamus caused the rat to lose many of its normal rhythms. Despite its smallness—in humans, the size of a pea—the hypothalamus is the Napoleon of brain tissue. It commands the sympathetic nervous system, where it speeds up one's breathing and heart in moments of danger, regulates body temperature, stimulates appetite and reproduction, and controls hormone secretion. After damage to the hypothalamus, Richter's rats slept, stayed awake, ate, drank, and ran in a seemingly normal manner, but at random times. Physicians have found that people who develop tumors that destroy their hypothalamus suffer the same fate; they awaken and sleep erratically, and they are no longer able to keep a regular schedule.

Richter's discovery of the hypothalamus as the key generator of body rhythms was akin to finding a needle in a haystack. Locating the critical region within the hypothalamus, the eye of the needle, so to speak, required still more sophisticated techniques for exploring brain anatomy. In 1972, two groups of scientists working independently, Friedrich Stephan and Irving Zucker of the University of California at Berkeley, and Robert Moore and Victor Eichler of the University of Chicago, recognizing the importance of light in setting rhythms, used radiographic tracers to follow the path of light from the eyes to the hypothalamus. They determined that the critical area of the hypothalamus was the *suprachiasmatic nucleus*, or SCN, a discrete, boomerang-shaped cluster of nerve cells about the size of a letter *v* on this page. With only eight thousand nerve cells, the SCN is a tiny island in the ten billion cells that make up the brain. It takes its name from its position on top of (the meaning of *supra*) the optic chiasm, a major junction for nerves from both eyes.

Researchers performed a variety of studies to confirm that the SCN was the body's master clock. They found that total destruction of the SCN abolished many rhythms of bodily functioning and behavior. (Some remain, for example, the rhythm of cortisol secretion. This hormone, important in metabolism and response to physical stress, is governed by the adrenal glands, located on top of the kidneys. Researchers do not yet know the relationship between the SCN and the adrenal clock.)

Researchers found that the SCN could keep time even when it was removed from the body. When SCN cells were placed in a laboratory dish with the proper nutrients, they continued to function with a circadian rhythm. Martin Ralph of the University of Virginia and his colleagues showed that when SCN tissue from one animal was transplanted into another whose SCN had been destroyed, it restored the recipient animal's circadian rhythms. Ralph later found that transplanting the SCN from one strain of hamster could reset the rhythms of a different strain; when an SCN from an animal with a twenty-hour cycle, for example, was transplanted into an animal that had followed a twenty-four-hour cycle before its SCN was destroyed, the recipient developed cycles similar to those of the donor.

As chief conductor of the body's biologic orchestra, the SCN directs the body's various rhythms to work together like professional musicians: ideally they will play their own parts well, and stay perfectly in tune with other performers.

The SCN treats daylight signals picked up by the eyes as a kind of musical score, enabling it to bid each body rhythm to play its part at the appropriate time, day after day. The SCN is a true maestro. It knows the score so well that even in the absence of daylight signals, it is able to tell the body rhythms what to do.

The SCN requires fine-tuning in the early days of life. "It would be disadvantageous to have inflexible rhythms in place on Day 1," explained Melanie Kittrell of the University of Pittsburgh. "The baby has to be ready to eat when its mother is available to feed it." Kittrell has found that rat pups do not develop strong circadian rhythms until they are three to four weeks old, the same time that they are weaned and become more independent.

Human circadian rhythms take somewhat longer to become fully entrenched. A regular twenty-four-hour temperature cycle appears in a normal full-term infant between six and eight weeks of age. Nonetheless, circadian tendencies manifest themselves quite early. For the first day or two after birth, babies wake and sleep in approximately equal amounts around the clock. But as soon as the third day, most babies sleep longer at night.

"At first mothers are forced to live on the baby's schedule," Kittrell noted. "But consciously or unconsciously, mothers try to move the baby to their schedule," she said. "Most parents will give a baby who is asleep at night every chance to stay asleep. Yet they don't hesitate to wake the baby after two hours or so of sleep during the day." Parents thus reinforce the child's awareness of the day as the right time to be active. As life proceeds, many other cues will combine to solidify daily rhythms, and enable the child to live comfortably in the twenty-four-hour world.

TIME CUES IN THE OUTSIDE WORLD

If plants and animals did not reset their slightly irregular internal clocks to a twenty-four-hour schedule every day, havoc would ensue. The early bird might be searching for the worm long before, or after, it has come and gone. The resetting process, called *entrainment*, is accomplished automatically with the aid of environmental cues. These are called *zeitgebers*, from two German words meaning "time-givers." The daily changes between light and dark—dawn and dusk—serve as the dominant zeitgebers for humans as well as for other organisms.

Until recently, many scientists believed humans had evolved without the need to rely on the sun to tell time. "One of the most important differences between man and apes and monkeys is that man, exposed as he has been for several hundred thousand years to artificial light at night, through his use of fire, has been able to free himself from the twenty-four-hour, and possibly other clocks—which in apes and lower animals are rigidly adjusted to the light of day," Richter concluded. He and his contemporaries thought that social behavior, such as family mealtimes, had supplanted light in organizing the human day.

The early studies of humans in the 1960s in laboratories shielded from time cues lent credence to the importance of social zeitgebers. Researchers at the Max-Planck Institute in Munich, Germany, controlled room lights to create days that were longer or shorter than twenty-four hours. Their subjects appeared to follow the schedules set by the room lights. One day, however, a gong system that was used to awaken the subjects, and signal them to provide urine samples, broke down. The subjects soon drifted out of sync with the lights.

The researchers surmised that the subjects had viewed the gongs as personal calls, and that this modest social interaction, not the lights, had been the most potent force structuring the subjects' days. But they were wrong. In 1980, researchers at the National Institute of Mental Health determined that the lights used in the early studies had simply not been bright enough.

Alfred Lewy and his colleagues showed that it took lights that were five to ten times brighter than ordinary room lights—the equivalent of daylight just after dawn—to reset human rhythms. They found that such lights were more potent than any other cue. Could human evolution have anticipated the invention of the light bulb? One might suspect that was so, since people were protected from having to adjust their rhythms whenever room lights were flicked on and off. Their body clocks somehow have learned to ignore light that is dimmer than daylight. But humans, like other animals, remain sensitive to the natural light/dark cycle.

Social zeitgebers, however, assuredly play a key role in organizing the human day. Perhaps the most important social cue in modern society is the ringing of the morning alarm clock. Its power comes not just from its ability to rouse one from sleep, but from what the sound signifies. The clock is a reminder of the penalties one may suffer for not getting to work or school on time. Knowledge of the time of day is a weak zeitgeber when those penalties are absent—on a day off, for instance, when the clock serves merely to remind one to do household chores. And if a clock serves no real purpose, it may be ignored entirely. A man who volunteered to live alone in a cave for three months, in a study of body rhythms during isolation, took along a wristwatch, intending to use it to stay synchronized with the world outside. He soon found that he stayed up later and later at night, and lingered in bed later and later in the morning. Soon, he tossed the watch aside and went to bed when he felt sleepy.

Light and social cues often reinforce each other, as anyone who has struggled to get out of bed on a dark winter morning knows. It is much easier to respond to the alarm clock when it is bright outside. Even on vacation, one can stash clocks out of sight, but one cannot ignore the sun.

Most people utilize a huge number of social zeitgebers to organize the day. Sharing a meal, watching a television news show, driving in rush hour, reading one's mail—these activities all serve as zeitgebers. So important are social zeitgebers that in their absence, people are more likely to suffer depression and other health problems. The death of a spouse, for example, disrupts the survivor's habitual mealtime. He or she may not eat as much food, or the same kinds of food, and may lose weight. Other habits also may fall by the wayside; these include a regular bedtime, a daily walk, and visits with friends. While depression is often blamed for poor sleep, bad eating habits, and the failure of the recently bereaved to participate in ordinary activities, these events in fact may set off the depression, and keep it going.

Alterations in social zeitgebers also may play a role in inducing the malaise and irritability that sometimes accompany jet travel and shift-work schedule changes. They may even be a factor in postpartum depression, experienced by one in ten new mothers. After giving birth, women may be wise to return as soon as possible to having meals at the usual times, watching favorite television shows, and otherwise sticking to a regular schedule.

THE PROMISE OF CHRONOBIOLOGY

Advances in chronobiology promise that in the twenty-first century:

- People will choose their jobs, and organize their days, mindful of their differing skills at different hours.
- Workplace productivity will increase, and industrial accidents will decrease. Public health hazards of night work and shift work will drop.
- Light will become a virtual "wonder drug," and will be used to boost alertness, improve sleep, treat infertility, relieve depression, and more.
- Falling asleep while driving will become a rarity.
- The school day for high school and college students will start later and end later than it currently does. Fewer students will doze in class even during the most boring lecture, in a warm room, with the lights dimmed to show slides.
- Physicians will synchronize medications and other treatments with body rhythms to improve cure rates for cancer, and to enable people with arthritis

to feel less pain, people with diabetes to achieve better blood sugar control, and people with epilepsy to have fewer seizures. These are only a few examples.

- Reliable jet lag remedies will make vacations more enjoyable, and enable shuttling diplomats, executives, athletes, musicians, and others to perform anywhere at the peak of their skills, just as they would at home.
- Dieters will learn to eat only when hungry, and the incidence of obesity will decline.
- Artists and writers will tap their most creative times.
- Lovers will know their peak times of sexual arousal.

Pie-in-the-sky fantasies? On the contrary, for some people many of these forecasts have already come true. Most people, however, have not yet incorporated the new knowledge of inner clocks into their daily lives. Nor has society widely applied chronobiologic principles to the workplace, the classroom, or the physician's consulting room. This book aims to help change that. It reports the emergence of chronobiology as a scientific discipline, and shows how everyone could benefit from learning a new way to tell time: body time.

CHAPTER 2

LIFE WITHOUT TIME CUES

IMAGINE LIVING ALONE in a windowless apartment or cave for several weeks or months at a stretch, with no clock, no radio or television, no morning newspaper, no clue at all to indicate the time, the date, or even whether it was day or night. One could stay up as late as one wanted, go to bed when one wished, stay in bed as long as desired, and eat whenever hungry. The only determinant of any activity would be an inner voice saying, "It's time."

But how would one decide when the time was right?

On any given day, perhaps only a handful of people anywhere in the world face this quandary. Only about five hundred people have ever known this experience. While writing this book, I became one of them. I served as a volunteer subject in a research study at the Institute of Chronobiology at the New York Hospital-Cornell Medical Center in White Plains, New York, one of the small number of centers worldwide where scientists probe into the workings of body clocks without the constraints imposed by the outside world. Such laboratories are essential to the study of human body clocks, because it would be impossible to limit and control the zeitgebers to which people living in the ordinary world are exposed. In such laboratories, researchers can determine the effect on the body of different amounts of sleep, activity, exercise, medications, caffeine, alcohol, and more, and they can assess the impact of a change in one or more of these factors on the others. They also try to determine why inner clocks sometimes go awry.

My home for two weeks was a one-bedroom apartment in a sequestered corner of the otherwise bustling hospital. Air-conditioned, carpeted, and attractively furnished, with a stereo and exercise equipment, the apartment also had a rare asset: soundproofing so effective it blocked hallway conversations, traffic, and even the whine of nearby elevators, all potential time-of-day cues. Of course, it lacked one ordinarily desirable feature: a view.

Still, these surroundings were far more luxurious than the cold, damp caves that early researchers of biologic rhythms used for their studies, and where some occasionally still set up shop today. In 1938, Nathaniel Kleitman, the father of sleep research, spent a month with one of his students, Bruce Richardson, in Kentucky's Mammoth Cave, where the beds, desks, and meals were provided by the Mammoth Cave Hotel. The event, not surprisingly, attracted considerable media interest.

The aim of the study was to see if people could live on a twenty-eight-hour day. To keep this schedule, the researchers stayed awake for nineteen hours, then turned off the lights, and stayed in bed for nine hours. The darkness was absolute when the lights were off, and there was complete silence, Kleitman recalled. Richardson managed to adjust to the new schedule within a week. Kleitman never did.

Kleitman slept well only when the sleeping hours coincided with his customary sleeping time; at other times, he had trouble falling asleep, or woke up early, or both. Richardson was twenty-three years old then, and Kleitman, forty-three. The study was an early demonstration that body rhythms become less flexible as people grow older. Kleitman concluded from the experiment that "there is no foundation for assuming that some cosmic forces determine the twenty-four-hour rhythm." The adherence to a twenty-eight-hour schedule, however, kept him from discovering how the body clock would have run on its own.

This finding had to wait until 1961, when West German physiologist Jürgen Aschoff started to monitor human cycles in an underground bunker at the Max-Planck Institute in Munich. The bunker had two suites, each with a fourteen-foot square room with a bed, a desk and chair, a small kitchen, and a bathroom, but no clocks. Aschoff served as the first subject for his own studies, monitoring his own temperature, urine output, mood, mental acuity, physical dexterity, and sleeping and waking cycles for several weeks before enlisting other volunteers.

These early studies confirmed that humans, like other animals, had predictable internal circadian clocks. They studies identified numerous rhythms, and mapped their course. Further, they offered researchers the opportunity to assess the relationships of the various activities to each other, finding, for instance, that certain hormones were released only during sleep.

The first chronobiology laboratory in the United States opened in 1977 at Montefiore Hospital in New York City. Its founder, Elliot Weitzman, later moved to Cornell, where he started the Institute of Chronobiology, which he directed until his death in 1983. In the studies that preceded Weitzman's, subjects in time-free environments lived in isolation, but Weitzman argued that humans were social animals and required social interaction for their body rhythms to function normally. The introduction of staff members into the time-isolation laboratory made possible biological measurements and psychological observations that previous constraints had not allowed. Weitzman and his colleagues charted rhythms in hormone release, sleep architecture, and other bodily functions under free-running conditions and after a variety of schedule changes.

During my stay at Cornell, I became "FR42," free-running subject 42, the low number a testament to the time-consuming effort needed to study inner clocks. Charles Pollak, the institute's director, explained that the study in which I was to participate had "something to do with eating," and that at some point during the experiment, he would ask me to modify my behavior. His explanation was deliberately vague to keep me from guessing the study's purpose and possibly influencing its results.

LIVING ON BODY TIME

During the two-week period, I set my own schedule for eating, sleeping, working, and other activities under the watchful eyes of the researchers. Ceiling microphones let me talk at any time from anywhere in the apartment with the technical staff in the adjacent space-age control room. The technicians also could see me on camera while I was in the living room, where I spent most of my waking hours. They were a gregarious crew who worked irregular schedules themselves to keep me from guessing the time. They greeted me with "Good morning" and "Good night" at the right time of my day, while giving me no clue to what time it was on theirs.

The first day, I plunged into a writing assignment I had brought along. I found myself glancing occasionally at my watchless wrist, only to realize with a start that I was cut off from time. Every so often, about fifty times a day, although I didn't know that then, I received a summons to rate my alertness on a computer terminal by moving a cursor along a line from "very sleepy" to "very alert."

More tests were administered whenever I ate, used the bathroom, exercised, went to sleep, or got out of bed. Each activity was followed by a short quiz to measure alertness, moods, mental skills, and manual dexterity. On going to bed, I answered a questionnaire about how the day had gone, and on getting out of bed, another on how I had slept. Because moods, cognitive skills, and physical performance vary predictably over the course of the day, the researchers could use my scores on the different tests to tell what time it was on my body clock.

One computerized test, the E search, demanded an instant response to whether or not the letter E appeared in a series of strings of letters flashed on the computer screen. This test measures a skill needed in everyday life in proofreading or product checking. Most people perform these tasks better in the afternoon than at any other time of day.

A second test measured verbal reasoning. Here, the computer presented a sentence such as "M comes before C" followed by the letters MC or CM. I had to decide whether the statement was true or false. "M comes before C: CM," for instance, is false. And that is an easy one. "C does not come before M: MC" takes more thought. Most people perform this test best in the late morning; it is devilishly difficult when one is fatigued. A third test involved putting small pegs in holes as fast as possible, a measure of motor skills needed for such activities as typing and athletics. Most people perform this best in the afternoon.

When I wanted to eat, I entered "request meal" on the computer terminal, designating my choice of breakfast, lunch, dinner, or snack. My choice told the researchers what time of day I believed it to be. Thanks to advance queries about food preferences, familiar foods were all on hand, from favorite breakfast cereals to dinner entrees. The only restrictions: no caffeine or alcohol, as both may alter one's mood and performance skills.

During the entire study, I wore a rectal thermometer. Thin and virtually unnoticeable, the thermometer was tethered by a long cord to the wall, permitting me to walk about the apartment freely while it provided a minute-by-minute temperature reading, more than twenty thousand bits of data before the study was over. Earlier studies in time-isolation labs had established that ups and downs in temperature constituted a reliable marker of what time it was on the body clock.

Temperature is at its low point, about 97°F, between 4 A.M. and 6 A.M. It moves upward before one ordinarily awakens; it rises sharply through the morning and more gently through the afternoon. It dips perhaps one tenth of a degree at midday, rises again, and peaks at around 99°F between 7 P.M. and 8 P.M. It then drifts downward, dropping more precipitously once one is asleep. Thus, a person experiences the so-called "normal temperature" of 98.6°F only twice a day, when temperature rises and falls, and then only for a few minutes. (Recent studies, using more accurate thermometers than those of the original investigators, put the daily average at 98.2°F. See chart, "The Daily Temperature Cycle," page 35.)

I hoped to complete a Herculean amount of writing while in the lab, a goal that prompted a routine similar to that of my life on the outside. I spent most of my "day" at work, rode an Exercycle after finishing, and then had dinner. After dinner, I read books, wrote letters, and listened to music on cassette tapes. I ate three meals a day, and usually had a bedtime snack.

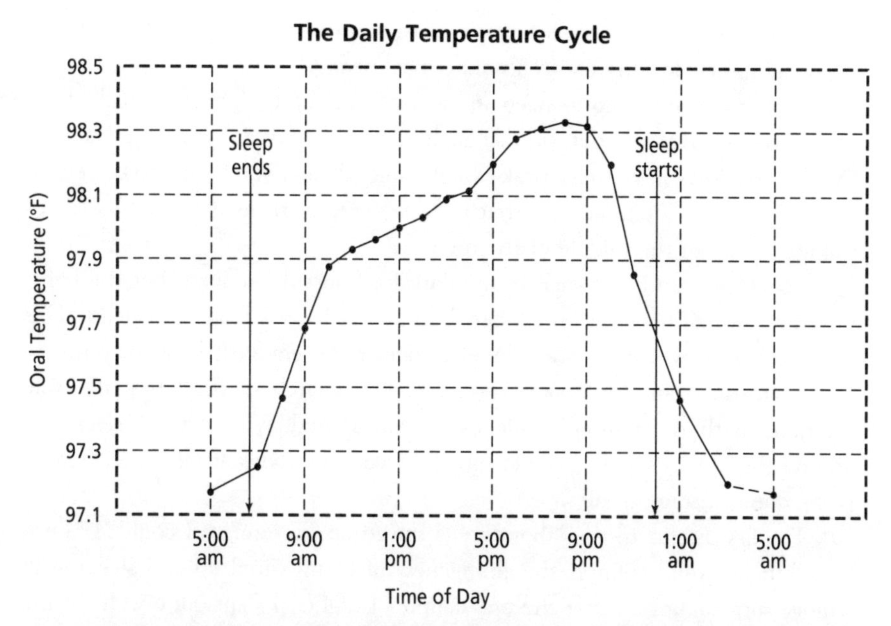

This chart shows the pattern of body temperature in young adults over the course of the day. Temperature is the touchstone circadian rhythm. It is easy to measure and highly predictable.

Data collected by W. P. Colquhoun. From Monk, Timothy. *How to Make Shiftwork Safe & Productive.* Des Plaines, Ill.: American Society of Safety Engineers, 1988. Reprinted with permission.

Sticking to this routine provided a rough way to mark time, of course. Without consulting a watch one knows approximately how long it takes to read a certain number of pages or run a mile, even on a jogging machine. And yet by the second day, I began to lose my certainty about what time it was in the outside world. Occasionally I asked the technicians to record the time I thought it was, along with the actual time, for my review after the study ended. I might as well have been using one of Salvador Dalí's wet watches. But even someone who attempted to measure time in a time-free environment, by playing thirty-minute tapes over and over, for example, would be thrown off by every episode of sleep.

BEDTIME—A TOUGH DECISION

When I announced that I had decided to go to bed, a technician would attach a dozen electrodes to my scalp, forehead, chin, and outside corners of my eyes to pick up the electrical activity of my brain and muscles while I slept. As soon as I got into bed, another technician in the control room would turn off the overhead lights, leaving me in absolute blackness.

Deciding when to go to sleep and when to get up took more thought than I had anticipated. I found it surprisingly hard to know when I was sleepy enough so that I wouldn't have to lie awake a long time. It was even harder to be certain on awakening in total darkness that I was ready to start my day, that I was not waking briefly in the middle of the night.

I could stay in bed awake in the dark for hours if I wished, but the lights would be turned back on only when I announced I was ready to get up. This decision was irrevocable. Once the electrodes were removed, I was supposed to stay up through another waking cycle. Some researchers assert that this rule interferes with the natural tendency to nap at midday. To assess sleepiness around the clock, subjects would have to wear electrodes at all times. This is being done in some studies.

The first nights in the laboratory, I stayed up "as late" as I could. I hoped that fatigue would counter the strangeness of being wired up and sleeping in strange surroundings. After the first couple of nights, I slept quite well. In the mornings, I savored the pleasure of staying in bed much later than usual, or so I thought. At home I hate to get out of bed. I seldom feel I have slept long enough. In the chrono lab, I awakened so alert and energetic, I assumed I was sleeping nine or ten hours instead of my usual seven. It was a real surprise to learn later that my longest sleep period lasted less than eight hours, and that I typically slept less than seven hours.

"Time isolation is probably the best and most legitimate way to tell how much sleep a person needs," Charles Pollak told me. "When people are allowed to sleep as long as they can," he said, "it is virtually impossible for them not to get all the sleep they need." In time isolation, as well as in everyday life, most adults spend about two thirds of their time awake and about one third asleep.

Indeed, people behave so predictably in time isolation that it is obvious that everyone possesses the innate ability to read his or her own built-in body clock. During the first part of my stay in the lab, I unknowingly went to bed and awakened each day about an hour later than the day before, as is typical for people who free-run. I was living on twenty-five-hour days. (See chart "Waking and Sleeping in the Chrono Lab," page 39.)

THE TWENTY-FIVE-HOUR CLOCK

The discovery of the twenty-five-hour clock has made it possible for researchers to solve numerous everyday mysteries. Why it is easier to stay up late than to go to sleep early is one example. As a practical illustration, most people who ordinarily retire at midnight find it easy to stay up until 1 A.M. This behavior conforms to the body's natural preference to lengthen the day; it is like slipping on a pair of loose shoes. By contrast, attempting to go to sleep at 11 P.M. would be harder, akin to squeezing into shoes that are too small.

Similarly, the reason that jet lag is felt more intensely after flying east than flying west, that shift workers have more trouble rotating counterclockwise than clockwise, and that the switch to daylight savings time in the spring is a bit more discomforting than the switch back to standard time in the fall is that these changes compress the day.

Suppose a person goes to bed at midnight in New York, then on the next day flies west to Denver across two time zones and goes to bed at midnight. That person will have experienced a twenty-six-hour, or loose-shoe, day. On the return trip, the same schedule will produce a twenty-two-hour, or tight-shoe, day. In chronobiology laboratory studies, researchers have asked subjects to live on days both longer and shorter than twenty-four hours. They have found that most people can adapt comfortably to days that are about two hours longer or shorter than twenty-five hours. This period, called the *range of entrainment*, is thus roughly twenty-three to twenty-seven hours long.

Most people will find a change to a schedule outside of the range of entrainment somewhat discomforting, as Kleitman did in his twenty-eight-hour-day experiment. A major thrust of contemporary research has been to develop ways of overcoming this problem, to the benefit of millions of jet travelers and shift workers. For example, researchers have found that subjects who were exposed at certain times of day to bright lights, with an intensity comparable to sunlight just after dawn, could live on days as short as nineteen hours or as long as thirty-one hours without experiencing difficulties with sleep or alertness. The new techniques could not have been devised without the knowledge gained in time-isolation laboratories.

THE LONGEST DAYS

At Cornell, after the study had been in progress for about a week, Charles Pollak requested that I try to stay up as late as possible. To comply, I decided that a good way to stave off sleep would be to prolong the intervals between meals. I had to ignore my body's hunger clock.

For a couple of days, I suffered all the ill effects of *circadian desynchronosis*, better known as jet lag. Without knowing why, I spent part of the day—the time I previously had been asleep—slumped over my desk, barely able to keep my eyes open. I struggled to concentrate. I felt chilly, achy, and cranky, much as if I were developing the flu. I became much more aware of the alerting value of social interaction and consciously sought the company of the chrono lab technicians to help me stay awake.

Although I felt miserable, I performed as well as previously on the various computer tests. This is not unusual in daily life. Even when people miss several hours of sleep, they manage most of their ordinary tasks about as well as they do customarily—from talking on the phone, to handling routine paperwork, to preparing a meal. Creative thinking, however, may suffer. The impact of sleep loss is discussed more fully in Chapter 4.

Once the period of wake-time lethargy passed, my alertness returned. Sometimes I would be sitting at the desk and suddenly feel as if I had just awakened from sleep and were ready to start a new day. At the end of these long days, I could answer without hesitation the question I had posed to myself earlier in the study: "Am I really ready to go to sleep now?" I was overwhelmed by sleepiness, so much so that one night (I later learned) I fell asleep three minutes after the light was turned off, far faster than is customary for me. My bedtime was pushed off progressively each day, traveling one and one quarter times around the clock. In fourteen days, I slept only twelve times.

After the study was over, I found out what Pollak was after: to split the rhythms of the body-temperature cycle and the sleep/wake cycle. Ordinarily, the two cycles are linked, with body temperature at its lowest during sleep. Occasionally, during time-isolation studies, for reasons that are still unknown, the two rhythms separate. In this situation, which is called *internal desynchronization*, body temperature may reach its low point during waking hours, a situation that has important consequences for both mental and physical performance, and for sleep.

In general, alertness is up when body temperature is up, and it is down when temperature is down. Sleep is restful when temperature is at its lowest, and it is hard to sustain when temperature is at its highest. When temperature is lowest during waking hours, alertness and dexterity sag. In my case, when the two rhythms split apart, my temperature continued to follow a twenty-five-hour rhythm, but my cycle of waking and sleeping stretched to more than thirty-one hours. As a result, my lowest temperature usually occurred earlier and earlier in my waking day, accounting for the malaise I experienced. (See chart, "Waking and Sleeping in the Chrono Lab," page 39.)

If scientists had a reliable way of separating these two important rhythms, Pollak said, they could learn more about which one controls behaviors such as

Waking and Sleeping in the Chrono Lab

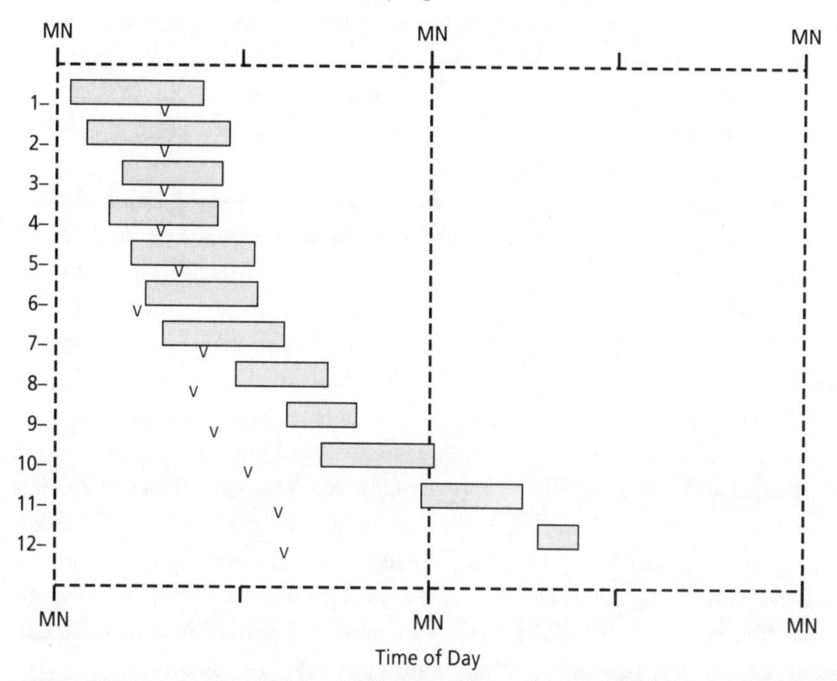

In the lab, the author spontaneously went to bed and awakened about an hour later each day, living on twenty-five-hour days, as is normal for people who are free-running. After one week, she was asked to try to stay up as late as possible. Her days stretched to thirty-one hours. The rectangles in the chart represent sleep episodes. Each V marks the low point of her daily body temperature. On days when her temperature lows occurred during waking hours, she suffered symptoms of jet lag.

eating, problem solving, and performance of motor tasks. (For more on the relationship of temperature and different types of mental performance, see Chapter 3, and for the relationship between temperature and sleep, see Chapter 4.)

During internal desynchronization in someone who is free-running, temperature will continue to have a twenty-five-hour cycle, but the rest/activity cycle may lengthen considerably, sometimes startlingly so. In 1988, Veronique Le Guen, a thirty-two-year-old Frenchwoman, spent 111 days in a cave in southern France, where she kept track of her own biologic rhythms. Le Guen once stayed awake for a double day of fifty hours, then slept for thirty hours, unaware that anything unusual had occurred. Another time, she felt drowsy after lunch and took what she called a "little" nap that lasted eighteen hours. When she awakened, she thought she had been asleep just a few minutes.

She stayed in her cave through three menstrual cycles, ordinarily a reliable marker of the passage of time. The cycles she experienced in the cave occurred at about the same frequency as her usual cycles, roughly four weeks apart. But because she was living on very long days, she thought her periods were coming as often as every eleven days. "If I had trusted my body," she said later, "I would have been very close to the real time."

One chronobiology laboratory study found that the temperature low point in a man who had been free-running for six months varied by less than ten minutes a day, while his cycle of sleeping and waking, like Le Guen's, ranged from twenty-five to fifty hours.

When people live on long days without being aware that they are doing so, one curious finding is that they still eat only their usual three meals. They do not increase the portion size, and yet, remarkably, they do not lose weight.

This implies, Pollak said, that under certain conditions, people use calories more efficiently; their metabolic rate actually may change. Studies in rodents show that changes in blood glucose patterns predict changes in eating behavior. Rodents start feeding when their blood glucose falls, then rises, he noted. Similar changes in humans may trigger the desire for a meal. "If we could identify the signal that prompts eating," he said, "we might be able to develop better treatment for a variety of weight problems, from starvation to obesity."

Participation in the study left me with an enhanced appreciation of the rigors of research, and of the collaborative effort required from both subjects and researchers to advance understanding of the human biologic clock. My brief encounter with simulated jet lag highlighted the dissonance one feels when body clocks run too fast or too slow, a problem that people encounter all too frequently nowadays. The sense of well-being I enjoyed in the first half of the study, however, made me long for clocks that always keep perfect time.

CHAPTER 3

DAY AND NIGHT

THE RHYTHMS OF ALERTNESS

ALMOST EVERYONE HAS HAD THE EXPERIENCE of staying up all night, fighting sleep through the predawn hours, and discovering that alertness returns in the morning. Some people attribute this phenomenon simply to willpower. But chronobiology studies demonstrate unequivocally that people who stay up all night will be more alert at 10 A.M. than they were at 4 A.M., even though they have gone an additional six hours without sleep. If they continue to stay awake, they will remain more alert throughout the day than they were the previous night. The longer they go without sleep, the sleepier they become, but they still will be more alert in the daytime than at night, and more alert at certain hours than at others. This rhythmic pattern of alertness and sleepiness is dictated by the biologic clock.

Early researchers in chronobiology theorized that alertness and body temperature went hand in hand, that the two rhythms peaked and fell simultaneously. Some researchers even claimed that one could assess mental fitness simply by taking a temperature reading. However, as researchers became increasingly adept at homing in on the disparate sensations and skills that made up the mental state called *alertness*, they found that the link between cognitive performance and body temperature was not that neat.

"How good people feel, how awake they feel, what they have to do, how well they think they can do it, how well they really can do it, how fast they can do it. . . . There is a long list of factors that people lump together but probably should not," according to Timothy Monk of the University of Pittsburgh. While many of these skills mirror the peaks and troughs in body temperature (see chart, "The Daily Temperature Cycle," page 35), some follow a different clock.

There is no single best time for all mental tasks but, rather, different peak times for different tasks. If it is crucial to be accurate, one time may be better than another. If it is crucial to be quick, another time may be appropriate. Studies of large numbers of people make it possible for scientists to predict when specific skills are likely to be optimal in about 80 percent of the population. In the remainder, extreme morning or extreme evening people, inner clocks may run markedly ahead or behind those of everyone else. (For more on these extremes, see "Larks and Owls," page 47.)

Variations in performance have obvious implications for both the classroom and the workplace. Over the normal waking day, one's cognitive performance efficiency may vary by as much as 15 percent above or below one's daily average, Monk has found. At their daily low of any particular mental task, he said, people may perform as poorly as they would after having three or four drinks or after missing all but three hours' sleep.

The executive who starts the morning's work by scanning the mail and getting routine correspondence "out of the way" may unwittingly forfeit the most creative part of the day. On the other hand, people may improve the quality of their work significantly by keying in to their own best and worst times and structuring their day accordingly.

ORGANIZING THE DAY

The following schedule shows how the typical dayworker may utilize the biologic clock to maximize performance.

7–8 A.M.: A regular time of arising is the single most effective way to keep body rhythms in tune. Exposure to sunlight within fifteen minutes of arising—or to artificial light of equivalent brightness—will aid further in anchoring rhythms. Even overcast winter days provide enough light. Exposure to bright light at any time of day may have an energizing effect, which implies that brighter lighting in schools, workplaces, and even homes might aid concentration and boost moods. Other methods of improving general alertness include exercise, which will raise body temperature, and eating some protein at breakfast.

Some people insist they need coffee, or another caffeinated drink, to get going in the morning. If they did not get enough sleep, that may be true. But caffeine often receives undeserved credit for the normal morning rise in alertness. If one is well rested, morning alertness will be at its peak, and caffeine will not provide any additional energy.

9 A.M.: The morning hours are prime time for organizing, planning, creative thinking, writing, and editing—tackling one's toughest projects.

10:30 A.M.: Contrary to popular belief, the traditional time for a break has no biologic imperative. The discovery during the 1950s that dreams occurred approximately every ninety minutes during sleep prompted the engaging notion that behavior was organized in ninety-minute patterns all around the clock. Studies were conducted on daydreaming, smoking, eating, and other voluntary behaviors, and researchers reported that they had identified numerous cycles that were approximately ninety minutes long.

Later studies, however, showed that the ninety-minute rhythms found in some waking activities were not synchronized with the ninety-minute pattern of dreams during sleep. Researchers also found that ninety-minute cycles were not as frequent as had been claimed, and that some of the initial wave of studies showing ninety-minute rhythms had been flawed. Regrettably, such research has done little to dispel the notion that brainstorms occur every ninety minutes.

These are the facts: When people are left in a room with food and drink and are told to help themselves, they will partake of the food and drink roughly every eighty to one hundred minutes. Most likely this is the result of stomach contractions, which themselves are caused by the cyclic release of digestive hormones. The initial smoking studies that showed a ninety-minute rhythm may simply have documented the association of smoking with eating. More recently, Daniel Kripke of the University of California at San Diego found that one-pack-a-day smokers habitually smoked at approximately twenty-minute intervals. Kripke and his colleagues also monitored general physical activity in normal adults who wore miniaturized computers on their wrists for seventy-four days while going about their everyday pursuits. No regular activity cycles were recorded.

11 A.M.: This is the best time to schedule a meeting. The majority of people rate themselves as most alert at this hour, and reasoning skills are at their daily high. Short-term memory, needed to keep facts readily accessible, is also at its peak; as a result, luncheon speakers, sales representatives about to make presentations, clergy heading for the pulpit, students about to take an exam, anyone who must have facts at his or her fingertips, will benefit from reviewing notes at this time.

Noon: Complex decision-making skills are at their peak.

1–2 P.M.: Cheerfulness is high. This is the best time for lunch.

Poultry makes a better business lunch than pasta, studies at the Massachusetts Institute of Technology (MIT) have suggested. Chicken, fish, and other foods high in protein may cause less of a postlunch slump than pasta and other high-carbohydrate foods. Protein is converted in the body to dopamine, a brain chemical associated with arousal, while carbohydrates prompt the manufacture of serotonin, a natural sedative. In the MIT study, Harris Lieberman and Bonnie Spring gave forty male students a luncheon sandwich of either turkey or a high-carbohydrate concoction containing the same number of calories. Those who had eaten the turkey performed faster on tests of coordination and intellectual skills throughout the afternoon than those who had eaten the high-carbohydrate lunch. "The differences were subtle," Lieberman said, "but they may be a factor in job performance."

The amount one eats also may influence the ability to concentrate. People make more mistakes on tasks that demand sustained attention, such as proofreading, after eating a 1,000-calorie meal than after a 300-calorie meal, according to research by Angus Craig of the University of Sussex in Brighton, England.

Having more or less food than usual also boosted errors, although those who ate large meals did worse as a group. Skipping lunch was not a solution. On an empty stomach, performance fell even lower. People who skipped a meal also felt more tense and anxious. Such findings give new meaning to the expression "power lunch."

2 P.M.: This is the time when daydreaming is most frequent. Radar operators, pilots, and others who must focus hard on their tasks may not be able to sustain concentration for as long as they could earlier in the day; thus they may benefit from scheduling more frequent breaks. Artists and writers, on the other hand, may find this a time of high creative energy that they would not want to miss by taking a long lunch break.

To study daydreams over a twenty-four-hour day, Leonard Giambra and his colleagues at the National Institute on Aging intermittently beeped volunteers while they were playing computer games. The subjects had to indicate if they had daydreamed since the last beep, and whether they had done so deliberately or whether the thoughts had occurred unbidden. The most frequent daydreaming occurred throughout the afternoon.

The researchers also found that daydream content had its own rhythm over the life cycle. With one exception, adults of all ages and both sexes daydreamed mainly about ways to solve personal problems. For men aged seventeen to twenty-nine, sex was the most popular subject. For everyone else, sex came second. After age forty, women daydreamed about achievements more often than men. After age fifty, both sexes had fewer daydreams with sexual themes.

2–3 P.M.: This period is often called the *post lunch dip*, although that is a misnomer. Most people experience transient sleepiness at this time regardless of when, or even whether, they have had lunch. Body temperature declines around this time by about one tenth of a degree, but that is not thought to be the trigger for the afternoon slump, since temperature remains higher than in the morning, when alertness was at its peak. The cause of the postlunch dip remains unknown.

This down time may be more pronounced after consuming carbohydrates or alcohol, following sleep deprivation, or during a boring task. To get through this period more easily in a typical office setting, one should interact with other people, make phone calls, run errands, or seek other activity.

This is a poor time of day for meetings. If one *must* make a presentation, adrenaline will flow and the performance should go well. However, the audience is likely to be drowsy, and more effort will be required to reach listeners than if one were giving the talk in the morning.

This also is a dangerous period to be on the road. Daytime single-vehicle accidents—most likely to be caused by lapses in attention—peak in midafternoon, especially in drivers over age forty-five. While there is no blood or breath test that confirms sleepiness in the same way tests of alcohol levels reveal that one has been drinking, sleep-related crashes have telltale signs: they lack skid marks or other indications that the driver tried to swerve, and injuries and damage to car and property are often severe. By contrast, when drivers who are drunk or are on drugs see that they are about to crash, they often jam on the brakes or whip the wheel around at the last minute. Such efforts sometimes reduce the devastation. Sleeping drivers literally never see it coming.

3 P.M.: The midday dip continues but should soon abate. Performance is high for mundane, repetitive tasks requiring motor activity, such as sorting mail. Typing speed increases over the course of the workday, but accuracy declines. Caffeine will have a more alerting effect at this time than at a morning break.

4 P.M.: Alertness returns. Now is a good time to tackle small projects, like writing letters, and to set priorities for the next day's tasks.

The typical office worker often must proofread text, verify order numbers or other figures, and watch out for various types of mistakes. Error detection is the prime task of the industrial quality-control inspector. Most people perform these tasks better after 4 P.M. and in the evening; the usual nine-to-five workday thus does not include the time when circadian performance is best.

Long-term memory, the skill needed to retain material over time, is highest in the afternoon. In a study conducted by Simon Folkard of the University of Sussex in England and his colleagues, a group of schoolchildren listened to a story at 9 A.M., while another group listened at 3 P.M. When tested right afterward to see how much they remembered, the children who had heard the story in the morning recalled the most. But a week later, the afternoon group scored higher. One implication of this finding for the workplace is that afternoon training sessions are likely to be more effective than those held in the morning. Similarly, students may retain a subject they study in the afternoon and early evening better than one they study in the morning. Because short-term memory is higher in the morning, however, the students may improve test scores by reviewing their notes early on the day of the exam.

Late-night cramming, incidentally, may benefit performance on an exam the next morning, Timothy Monk found. Unfortunately, people forget material learned between midnight and dawn more quickly than material learned at other times of day.

5 P.M.: This is a good time for exercise, particularly sports demanding eye-hand coordination, such as tennis, baseball, and archery. At this hour, people are most coordinated, their grip strength strongest, and their reaction time swiftest. Additionally, people who exercise for at least twenty minutes in the late afternoon fall asleep faster, and sleep better at night, than those who exercise at other times or not at all. Afternoon exercise causes an increase in body temperature that is followed six hours later by a decrease in temperature; when body temperature falls near bedtime, people usually become sleepy.

6–8 P.M.: Dinnertime. Evening plans may influence choice of foods. A meal that is high in carbohydrates may promote a more relaxing evening than one that is high in protein.

8–10 P.M.: Alertness remains high. This is a good time to pay bills and balance the checkbook.

A social drinker who expects to drive home from a party would be wiser to imbibe at the start of the evening than when it is winding down. That is because alcohol consumed during daily lows in alertness—which include the hours near bedtime and when one normally sleeps—will have a more potent effect than alcohol consumed earlier in the evening. The effect of alcohol on the body is also related to the amount of sleep one has had the night before, Timothy Roehrs and his colleagues at the Henry Ford Hospital in Detroit have found. Two to three drinks consumed after one has missed as little as three hours of sleep will have the same effect as five to six drinks when one is well rested. The effect of alcohol on the central nervous system outlives its effect in the blood, Roehrs found. The alcohol in two to four drinks may disappear from the bloodstream in four hours, but the sleepiness it induces may persist several hours after that. This explains why a person who drinks heavily one night may feel sluggish the next afternoon.

10–11 P.M. A bedtime snack to promote sleepiness is one high in carbohydrates: fruit juice plus a couple of cookies or a slice of cake, for example. Alcohol

SELF-TEST: ALERTNESS AND TEMPERATURE CYCLES

To plot your own alertness cycle, use the sleepiness/alertness chart on pages 234–235. Or draw a series of six-inch lines across a page. Label the left side "very sleepy" and the right "very alert." Run the hours down the left margin. Every hour or so during your waking day, put a mark somewhere along the line for that hour.

To plot your own body temperature, write the hours of the day down the left side of the page, and enter temperature readings from 96° to 100°F across the bottom. Shake an oral thermometer down below 95°F. Take readings every two hours, keeping the thermometer under your tongue for three minutes, and read it to the nearest tenth of a degree. Fifteen minutes before taking a reading, do not go outdoors, exercise, bathe, eat, or smoke.

is not a good soporific. It initially may induce feelings of sleepiness, but as it is metabolized, it will cause frequent awakenings.

Midnight: Going to bed at about the same time each night programs the body clock to expect sleep. It also reduces night-to-night variability, which has the psychological benefit of minimizing anxiety about sleep, perhaps the most common cause of insomnia.

LARKS AND OWLS

About one in ten people is a distinct morning person, or lark, and another one in ten is an evening person, or owl. Strictly speaking, there are no night people: even at the extremes, humans are diurnal creatures. Like having brown hair or blue eyes, lark and owl traits are inborn. Lark and owl traits have a potent impact on daily life: larks are the first ones at the office; owls are the last to leave. Larks exercise before they go to work; owls, afterward. Owls disdain breakfast; many larks say it is their favorite meal. Larks enjoy soft music more than owls, and they drink less alcohol. Under deadline pressure, larks get an early start, while owls work past midnight.

Artist Edward Hopper often focused on these two extremes. In "Cape Cod Morning," a woman, dressed for the day, looks out of her living room window at trees barely touched by dawn light. In "Nighthawks," a couple lingers over coffee at the counter of an all-night diner.

Socrates conducted his celebrated dialogues in the late afternoon and evening. René Descartes, the seventeenth-century French philosopher and mathematician, claimed he owed his best ideas to the fact that he never got out of bed before noon. Thomas Edison, an inveterate owl, earned the gratitude of owls everywhere by inventing the electric light. Many owls would agree with Jimmy Walker, mayor of New York in the 1920s, who once said, "It's a sin to go to bed the same day you get up."

The noted behaviorist B. F. Skinner, a lark, went to bed every night at 10 P.M. He slept three hours, then got up and worked for an hour at his desk. He then went back to bed for another three hours, rising at 5 A.M. to start his day.

Despite larks' and owls' often marked behavioral differences, their body temperatures are on virtually the same schedule as everyone else's. In larks, temperature peaks at about 7:30 P.M., and in owls, only about an hour later. Larks typically go to bed and get up more than ninety minutes earlier than owls. In one study, volunteers stayed in bed in the dark between 10 P.M. and 8 A.M. Although they could sleep whenever they wanted, the larks still went to sleep and awakened earlier than the owls. Both groups, however, got about the same amount of sleep. For tasks in which temperature and performance run parallel, owls lag about an hour behind larks. In studies of logical reasoning skills, Timothy Monk found that owls were at their best at 11 A.M., the same time as people with no decided morning or evening preference. Larks, however, were at their peak at 8 A.M., a difference that cannot be accounted for by temperature rhythms or sleep habits.

Personality tests fail to support the still widely held view that larks are more withdrawn and owls, more outgoing. "There is little by way of obvious personality differences between owls and larks," according to James Horne of Loughborough University in Leicestershire, England, codeveloper of the Owl/Lark Self-test. (See page 50.) "One can easily find the extroverted morning type who is the life and soul of the breakfast table, and the introverted evening type who reads well into the night," Horne said.

In a study comparing people who were up at night with others who were up during the day, however, sociologist Murray Melbin of Boston University found more individualism in nighttimers than in daytimers. Melbin views the world after dark as a frontier, and likens its colonists to those who settled the West. The adversity these people share by being deprived of activities and services available only to those who are awake during the day, he has suggested, breeds in them an extraordinary willingness to go out of their way for others, even in large cities, often portrayed as "centers of aloofness."

Melbin directed a series of experiments in Boston over all hours of the day, all days of the week. Researchers rated more than 2,500 people on how they behaved in public toward others during casual encounters that, according to Melbin, "exemplify the so-called little things that people often refer to when they talk about a humane quality of life."

Pairs of male and female researchers asked directions to a well-known location. They also asked passersby to answer a survey about people in cities. They observed cashiers and customers in twenty-four-hour supermarkets to see if people smiled and chatted about matters other than the transaction. In these face-to-face encounters, people proved to be more helpful and friendlier at night. Strangers provided directions agreeably. A few even offered the research couple a ride. More people answered survey questions. Sociability at the checkout counter, while rare, still was higher at night.

In another experiment, Melbin explored how people behaved when they found a key with an address tag lying in the street. About three hundred keys were picked up. Only half of the keys found at night were mailed back immediately, compared with two thirds of those found during the day. Even these results supported the pioneer mentality theory, Melbin said: because night people see fewer people around, they assume that the keys belong to daytime people. He suggested this showed they were more inclined to be friendly and helpful toward night owls like themselves.

WHEN LARKS AND OWLS MARRY

While most people express some preference for mornings or evenings, they can adjust their behavior as circumstances demand. The college student who seldom goes to sleep before 2 A.M. will usually be able to shift after graduation to the earlier schedule of the nine-to-five work world. The same person will be able to shift to an even earlier day after becoming a parent. People who are extreme owls or larks, however, may find it difficult, if not impossible, to modify their habits. That may cause problems in the workplace (see Chapter 10) as well as in marital and family life. Mismatched couples spend less time in serious conversation and shared activities, including sex, than those who are well matched.

In a survey of 150 middle-class couples, all similar in most ways, Brigham Young University psychologist Jeffry Larson found that more than half were out of sync. One in three couples in this group had troubled marriages, compared with only one in twelve of those who were more evenly matched.

"The challenge for mismatched couples is to create togetherness, while

THE OWL/LARK SELF-TEST*

When urged to select the best time of day for certain activities, people sometimes discover a preference that is in conflict with work or family responsibilities. The test below is designed to increase self-awareness. Some industrial psychologists have suggested that tests of "morningness" and "eveningness" would help employers to assess a prospective employee's suitability for specific work shifts. Their reasoning goes beyond improving job satisfaction. Workers who report poor performance at certain times of day, if assigned to those time slots, may pose a danger to themselves or others.

Instructions:

- Answer all questions in numerical order.
- Answer each question independently of others. Do not go back and check your answers.
- Select one answer only. Some questions have a scale. Place a mark at the appropriate point along the scale.

1. Considering only your own "feeling best" rhythm, at what time would you get up if you were entirely free to plan your day?

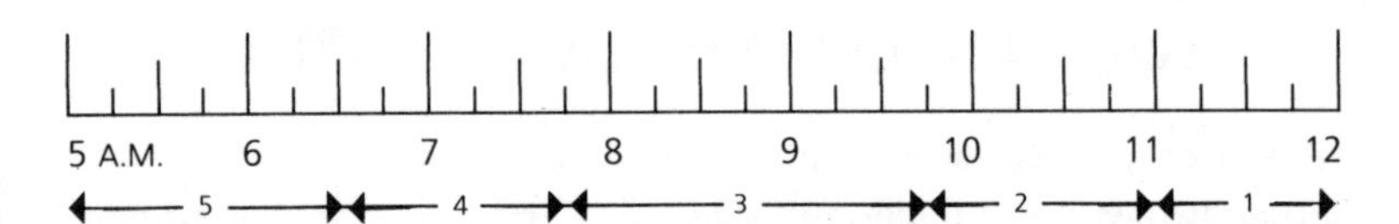

2. Considering only your own "feeling best" rhythm, at what time would you go to bed if you were entirely free to plan your evening?

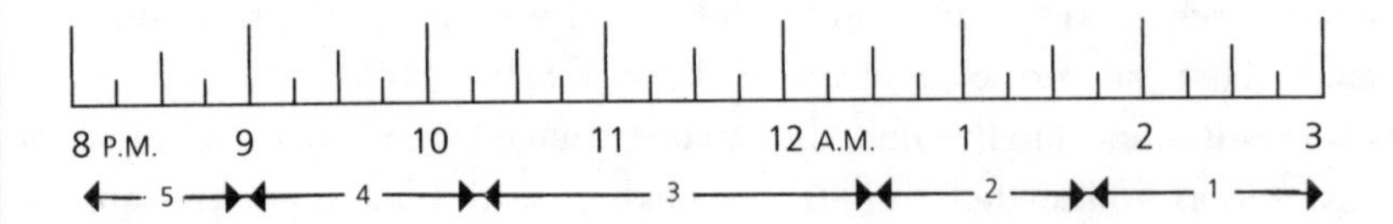

*Adapted from "A Self-Assessment Questionnaire to Determine Morningness-Eveningness in Human Circadian Rhythms," by James Horne and Olov Ostberg. *International Journal of Chronobiology*, London, England: Gordon and Breach Science Publishers Ltd., 1976;4:97–110. (Reprinted with permission.)

THE OWL/LARK SELF-TEST, *Continued*

Question	Answer	Score
3. If there is a specific time at which you have to get up in the morning, to what extent are you dependent on being woken up by an alarm clock?	Not at all dependent	☐ 4
	Slightly dependent	☐ 3
	Fairly dependent	☐ 2
	Very dependent	☐ 1
4. Assuming adequate environmental conditions, how easy do you find getting up in the mornings?	Not at all easy	☐ 1
	Not very easy	☐ 2
	Fairly easy	☐ 3
	Very easy	☐ 4
5. How alert do you feel during the first half hour after having woken in the mornings?	Not at all alert	☐ 1
	Slightly alert	☐ 2
	Fairly alert	☐ 3
	Very alert	☐ 4
6. How is your appetite during the first half hour after having woken in the mornings?	Very poor	☐ 1
	Fairly poor	☐ 2
	Fairly good	☐ 3
	Very good	☐ 4
7. During the first half hour after having woken in the morning, how tired do you feel?	Very tired	☐ 1
	Fairly tired	☐ 2
	Fairly refreshed	☐ 3
	Very refreshed	☐ 4
8. When you have no commitments the next day, at what time do you go to bed compared to your usual bedtime?	Seldom or never later	☐ 4
	Less than one hour later	☐ 3
	1–2 hours later	☐ 2
	More than two hours later	☐ 1
9. You have decided to engage in some physical exercise. A friend suggests that you do this one hour twice a week and the best time for him is between 7:00–8:00 A.M. Bearing in mind nothing else but your own "feeling best" rhythm, how do you think you would perform?	Would be in good form	☐ 4
	Would be in reasonable form	☐ 3
	Would find it difficult	☐ 2
	Would find it very difficult	☐ 1

THE OWL/LARK SELF-TEST, *Continued*

10. At what time in the evening do you feel tired and as a result in need of sleep?

8 P.M.	9	10	11	12 A.M.	1	2	3

← 5 → ← 4 → ← 3 → ← 2 → ← 1 →

11. You wish to be at your peak performance for a test which you know is going to be mentally exhausting and lasting for two hours. You are entirely free to plan your day and considering only your own "feeling best" rhythm, which ONE of the four testing times would you choose?

- 8:00–10:00 A.M. ☐ 6
- 11:00 A.M.–1:00 P.M. ☐ 4
- 3:00–5:00 P.M. ☐ 2
- 7:00–9:00 P.M. ☐ 0

12. If you went to bed at 11:00 P.M., at what level of tiredness would you be?

- Not at all tired ☐ 0
- A little tired ☐ 2
- Fairly tired ☐ 3
- Very tired ☐ 5

13. For some reason you have gone to bed several hours later than usual, but there is no need to get up at any particular time the next morning. Which ONE of the following events are you most likely to experience?

- Will wake up at usual time and will NOT fall asleep again ☐ 4
- Will wake up at usual time and will doze thereafter ☐ 3
- Will wake up at usual time but will fall asleep again ☐ 2
- Will NOT wake up until later than usual ☐ 1

14. One night you have to remain awake between 4:00–6:00 A.M. in order to carry out a night watch. You have no commitments the next day. Which ONE of the following alternatives will suit you best?

- Would NOT go to bed until watch was over ☐ 1
- Would take a nap before and sleep after ☐ 2
- Would take a good sleep before and nap after ☐ 3
- Would take ALL sleep before watch ☐ 4

THE OWL/LARK SELF-TEST, *Continued*

15. You have to do two hours of hard physical work. You are entirely free to plan your day and considering only your own "feeling best" rhythm, which ONE of the following times would you choose?
    - 8:00–10:00 A.M. .... ☐ 4
    - 11:00 A.M.–1:00 P.M. .... ☐ 3
    - 3:00–5:00 P.M. .... ☐ 2
    - 7:00–9:00 P.M. .... ☐ 1

16. You have decided to engage in hard physical exercise. A friend suggests that you do this for one hour twice a week and the best time for him is between 10:00–11:00 P.M. Bearing in mind nothing else but your own "feeling best" rhythm, how well do you think you would perform?
    - Would be in good form .... ☐ 1
    - Would be in reasonable form .... ☐ 2
    - Would find it difficult .... ☐ 3
    - Would find it very difficult .... ☐ 4

17. Suppose that you can choose your own work hours. Assume that you worked a FIVE-hour day (including breaks) and that your job was interesting and paid by results. Which FIVE CONSECUTIVE HOURS would you select?

12 1 2 3 4 5 6 7 8 9 10 11 12 1 2 3 4 5 6 7 8 9 10 11 12
MIDNIGHT NOON MIDNIGHT

←1→←5→ 4 ←3→←2→←1→

18. At what time of the day do you think that you reach your "feeling best" peak?

12 1 2 3 4 5 6 7 8 9 10 11 12 1 2 3 4 5 6 7 8 9 10 11 12
MIDNIGHT NOON MIDNIGHT

←1→←5→←4→←3→←2→←1→

THE OWL/LARK SELF-TEST, *Continued*

19. One hears about "morning" and "evening" types of people. Which ONE of these types do you consider yourself to be?
    - Definitely a "morning" type? .... ☐ 6
    - Rather more a "morning" than an "evening" type .... ☐ 4
    - Rather more an "evening" than a "morning" type .... ☐ 2
    - Definitely an "evening" type .... ☐ 0

Scoring:

- *The score for each response is beside the answer box or in a range below the scale. For question 17, use the most extreme mark on the right-hand side to find your score on the range below.*
- *Total your scores and compare them to the scale below.*

Definitely Morning Type	*70–86*
Moderately Morning Type	*59–69*
Neither Type	*42–58*
Moderately Evening Type	*31–41*
Definitely Evening Type	*16–30*

matched couples have to work to create distance in their relationship," sociologist Bert Adams, of the University of Wisconsin, and family therapist Ronald Cromwell, of the University of Missouri, concluded after surveying twenty-eight teachers about owl/lark issues in their marriages. Although matched couples in their study reported fewer conflicts, mismatched couples who make a go of it, the researchers said, "are likely to exhibit greater flexibility and adaptability in various aspects of their relationship than matched couples."

A mismatch occasionally may prove to be an asset. After the birth of a baby, an owl may handle night duty. As children grow, the lark may be the one to see them off to school.

HOW CAN ONE IMPROVE ALERTNESS?

Caffeine may improve alertness, alcohol may lower it, and certain foods may have either a stimulant or sedating effect, but none has as great an impact on daily life as the adequacy of one's sleep. Yet, according to the National Commission on Sleep Disorders Research, too many Americans get less sleep than they need to achieve optimal alertness. Objective tests show that the average adult is only slightly more sleepy than people with medical problems that compromise alertness, such as sleep apnea, a disorder in which breathing stops repeatedly during sleep, and narcolepsy, whose most prominent symptom is sleepiness. As many as one third of all American adults—forty million people—suffer from some disorder or disturbance of sleep, the commission said in its 1993 report to the U.S. Congress. Moreover, America faces an ever-expanding national sleep debt.

The need for sleep is as basic as those for food and water. Although much about it remains a mystery, sleep is no longer the locked and chained black box that it once was. Discoveries made since the 1950s have helped to pry the lid open.

CHAPTER 4

NIGHT AND DAY

THE RHYTHMS OF SLEEP

SLEEP IS A TIME OF VAST POTENTIAL DANGER. Lions and lambs alike must find a safe place to lie down. They cannot protect themselves from predators while sleeping, and they cannot protect their young. Early humans, living in caves or huts, slept facing the entrance, the better to ward off an attack. Even today, in houses with secure locks, many people refuse to sleep with their head nearer to the door than their feet. Although the sense of vulnerability persists, all mammals sleep, and they do so every day. What prompts such risky behavior?

Research conducted at the University of Chicago over the past decade suggests that one major purpose of sleep—perhaps the most important purpose—is the conservation of energy. During sleep, the body turns down its thermostat, thereby lowering its need for food. The body's metabolic rate, an indicator of energy expenditure, may run at only three quarters of its waking speed. Staying awake all the time would not only require eating more food but also more time to find the food and prepare it. Both food supplies and the quality of life would rapidly diminish.

Sleep also may be necessary for the restoration of energy. Shakespeare addressed this notion when he called sleep "nature's soft nurse," and it is amply supported by everyday experience. Nonetheless, scientists have been unsuccessful in their search for a specific substance that either is depleted during waking and built up by sleep or, conversely, accumulates during waking and is dissipated by sleep. However, they have found that cell division, protein synthesis, increases in sex hormones at puberty, and other growth processes occur at higher rates during sleep than when one is awake. "If sleep does not serve an absolutely vital function," University of Chicago researcher Allan Rechtschaffen has said, "then it is the biggest mistake the evolutionary process has ever made."

Rechtschaffen's research has shown that rats deprived of sleep sickened and died in eleven to thirty-two days. By contrast, control rats treated exactly the same way, but permitted to sleep, remained perfectly healthy. While sleep deprivation has dire effects, "we have a miracle cure," Rechtschaffen has asserted. "It's sleep." When rats that showed the ill effects of sleep deprivation were allowed to sleep, they recovered completely.

The documented record for humans for staving off sleep belongs to Randy Gardner, a seventeen-year-old high school senior. Gardner blazed his way into *The Guinness Book of Records* by staying awake for 264 hours and 12 minutes, an amazing 11 days, for a school science-fair project in 1964. (The previous world record was 260 hours.) Although somewhat irritable after about the fourth day of his adventure and increasingly drowsy, Gardner kept active. After nearly ten sleepless days, he stumbled through the alphabet but still managed to win games of pinball. Once he surpassed the previous world record, he fielded questions at a press conference, then went to bed and fell asleep within two minutes. He awakened fourteen hours and forty minutes later, feeling fine.

Such stunts aside, people rarely attempt to go without sleep for long. There is only a handful of human studies on the impact of going without sleep for more than five days. No one has ever suffered lasting ill effects, although participants have shown a decline in their ability to concentrate and to remember information, as well as a decrease in their reaction time. They have also shown disturbances of mood, and in some cases have experienced hallucinations and delusions—strong evidence that sleep benefits mental functions. The common expression "to sleep on it" testifies to the widespread belief in the utility of sleep for problem-solving. Indeed, cognitive activities, such as decision-making, mathematic calculations, and pattern recognition, show more serious decrements after sleep loss than do physical activities, such as marching.

In a study at Loughborough University in Leicestershire, England, psychologist James Horne gave twenty-four college students a series of sophisticated tests that required creative thinking. One test challenged the students to come up with potential uses for a cardboard box, or to imagine the consequences of a bizarre event—what would happen, for instance, if clouds had string attached to them hanging down to earth.

Initially, all the students had comparable scores in creativity and in problem-solving ability. Then, half the students were asked to stay awake all night. The next day, the sleep-deprived students scored one third to two thirds lower in the same areas. Moreover, they kept trying to solve new problems with approaches that had worked previously, even after it had become obvious that those same strategies were ineffective.

IMPACT OF SLEEP LOSS ON DOCTORS IN TRAINING

In 1988, the *Journal of the American Medical Association (JAMA)* published an unsigned essay titled "It's over, Debbie," in which a gynecology resident wrote about receiving calls during the night. "I had come to detest telephone calls," the author said, "because invariably I would be up for several hours and would not feel good the next day." The resident wrote that a desire to get back to sleep in the middle of the night prompted him (or her) to administer a lethal dose of morphine to a twenty-year-old woman. In pain and dying from cancer, the patient had said only, "Let's get this over with," and further, she was a patient the resident had never seen before.

This essay understandably sparked a major furor, most of it focusing on the question of euthanasia. In a letter to the editor of *JAMA*, William Fiorini, a physician in Somerset, Ohio, eloquently addressed the soundness of middle-of-the-night decision making: "Too often in medicine those challenges requiring the most careful consideration occur when the fog of sleep deprivation and overwork clouds the very sense and sensibility demanded for a proper response."

Medical training has traditionally demanded long hours with little sleep, and residents have commonly worked shifts of 36 hours per day, and 125 hours per week. Some physicians claim that arduous training toughens young doctors for a lifetime of being constantly on call, that it is a professional rite of passage. They say that only by providing round-the-clock care to acutely ill patients can students learn how a disease progresses, and how to think and act under pressure. It is also claimed that the patients benefit from the continuity of care.

A more jaundiced view is that residents provide cheap labor for hospitals. Along with direct patient care, residents transport patients, draw blood, take specimens to laboratories, and track down charts and X-ray films. Indeed, the term "resident" is a carryover from the days when physicians in training lived at the hospital all the time. British researchers compared residents' work schedules with those mandated for British pilots. If residents worked within the same guidelines, they said, twenty-six of them would have been needed to cover the hours that six actually spent on duty.

Today's doctors receive their training in a high-tech, high-stress environment with many more demands than were made on their predecessors. Because of pressures to keep costs down, people admitted to hospitals are sicker and they are out the door quicker than in days past—a situation that, in fact, deprives residents of the opportunity to follow their patients' progress. Because the population is aging, residents see more elderly people, who are more likely to have multiple disorders. They also see more drug users and homeless people, most of whom are seriously ill before they enter treatment. They must deal with AIDS, a complex and fatal disease virtually unknown a decade ago, which brings the added stresses of seeing younger patients die, and they must deal with the concern of possibly becoming infected themselves. Residents now manage illnesses routinely on regular patient floors that until recently were treated only in intensive-care units. They must familiarize themselves with hundreds of medications and laboratory tests. And they have much more paperwork to do.

Indeed, house physicians on call may spend more time on the care of their charts than on the care of their patients. University of Minnesota researchers found that interns spent more time writing up the results of a new patient's history and physical exam than they did in taking the history or performing the exam. Nicole Lurie and her colleagues learned that young doctors sometimes received two dozen calls a night. Most calls were to draw blood or perform other tasks that technicians or other nonphysicians could have handled competently. Visits to patients were interrupted every few minutes. The doctors spent little time—two to five minutes a night on average—talking in person with patients' families. Moreover, they seldom got even a single hour's sleep without interruptions.

Of hundreds of studies on sleep deprivation, fewer than two dozen studies focus specifically on residents; perhaps residents are simply too busy. Some studies were flawed: they recorded data on residents before and after a thirty-six-hour shift without taking the residents' chronic sleep deprivation into account. The most widespread finding is that sleep deprivation may lessen compassion, prompting doctors to refer to patients in derogatory terms, such as "crock" or "gomer," the latter meaning "get out of my emergency room."

The situation described in "It's over, Debbie" is an extreme one, perhaps more fiction than fact (*JAMA* has never indicated). Sleep specialists say that sleep deprivation has a greater impact on routine tasks than on extraordinary ones. "We all can alert ourselves for short periods in a crisis," said Timothy Roehrs of the Henry Ford Hospital in Detroit. "As a result, we may fail to appreciate how sleepy we are. After a night of heroic efforts to save a patient's life, a sleepy resident may neglect to write in the patient's chart some critical piece of information that might affect later care. Sleepy residents may miscalculate a drug dose. They may accept someone else's interpretation of an x-ray, instead of going to look for themselves."

Pediatrician and author Perri Klass was sitting at the nurses' station at 4 A.M. writing a note on one of her patients. A radio was playing softly. "My neurons crossed, and instead of writing out the patient's history, I began to transcribe the words of the song," she recalled. " 'The patient was in his usual state of good health until two days prior to admission when he developed fever and headache and I wish they all could be California girls,' " she wrote in the chart. "I sat staring down at the paper, trying to figure out what was wrong with it," she reported. "It made perfect sense to me."

The personal lives of harried residents also suffer. Doctors who work 125 hours a week have just 43 hours a week—or about 6 hours a day—to sleep, eat, shower, and commute to and from work, let alone talk to their families or pay their bills. Two in five residents in one survey reported that their work performance suffered because of marital problems. A study of pregnancy outcomes found that female residents had voluntary abortions three times more frequently than wives of male residents. One in three young doctors in another study reported "significant" depression, a much higher rate than in the rest of the population. And another study found that residents had three times more automobile accidents or near-misses during their residency than they had before entering the training program.

"GETTING BY" VERSUS "FEELING FINE"

Residency programs may cause severe chronic sleep deprivation; other situations may make excessive short-term demands on the body's sleep system. Under stressful and continuous conditions, such as war or earthquake rescue missions, or in extreme situations, such as space travel, sleep is typically brief and fragmented. "The need for sleep probably is the most important limitation during prolonged periods of continuous work," according to Claudio Stampi of the Institute for Circadian Physiology in Boston. If people do not manage their sleep correctly, they tend to go totally without sleep or to keep going on minimal sleep, he said. They build up a sleep debt, and as a result, they often sabotage their ability to perform efficiently.

Many sleep-deprivation studies have been conducted of military troops in situations mimicking field conditions: deployment on short notice and going for long stretches without rest. Navy studies, for example, showed that after two to three days without stop, performance of physical and mental tasks became unacceptably poor. In a British army trial lasting nine days, one group of soldiers got only ninety minutes of sleep per twenty-four hours, while another group was allowed three hours. Only half of the first group completed the exercise, but nearly all the soldiers in the second group made it to the last day.

Yachtsmen participating in solo races across the oceans provide a nearly perfect model for studying how people structure their lives when they must work around the clock, Stampi said. During a race, the solo yachtsman faces recurrent and unpredictable demands to stay awake for several days or weeks. He or she must steer and adjust the yacht under different, and sometimes threatening, wind and sea conditions. The person also has to keep up with what competitors are doing, and try to outfox them. While he or she grabs some sleep, the yacht may go off course or fall behind. Moreover, even with normal visibility, a yachtsman must check the horizon every fifteen minutes to avoid colliding with another ship.

In studying more than two hundred solo sailors, in transatlantic crossings as well as shorter races, Stampi found that the top racers slept often, but typically for only ten to sixty minutes at a time. They got by on about four to five hours of sleep a day compared with their usual seven hours while on land. The bottom line is that in the short run—for a few days or even weeks, although probably not for the three or more years required by most medical training programs—people can perform most tasks about as well as they ordinarily would on 60 to 70 percent of their usual daily amount of sleep.

Such findings are fostering the development of schedules designed to squeeze the most benefit from the smallest amount of sleep. Leonardo da Vinci's sleep habits may be worth emulating, Stampi has suggested. Legend holds that Leonardo sought to enhance his productivity by sleeping for only fifteen minutes every four hours, a total of ninety minutes a day. Under Stampi's direction, a twenty-seven-year-old graphic artist managed to reduce his sleep over the course of one week from about seven hours a day to less than three hours a day. The artist continued to work, taking six naps a day during the three-week experiment. "He was sleepy," Stampi conceded, "but he felt the trade-off was worth it."

In a study at Boston's Brigham and Women's Hospital, Gary Richardson and his colleagues found that a little sleep could make a big difference in the performance of resident physicians. The researchers compared two pairs of young doctors: one member of each pair had a protected sleep period during the on-call night, from either 11 P.M. to 3 A.M., or from 3 to 7 A.M.; the other took calls all through the night as usual. Those doctors who got three or four hours' sleep far outperformed their peers. (For more information on the most effective sleep schedules at times when one must miss some sleep, see the Appendix, "Strategies for Self-help.")

The flip side of trying to find out how little sleep people can get by on is to find out how much they need to function at their best. Roehrs and his colleagues have shown that people benefit from sleeping as long as they can. Previous research had suggested that oversleeping causes grogginess, much as overeating causes weight gain, but Roehrs's group found that that is not the case. The researchers enlisted twenty-four healthy men who normally slept between six and eight hours a night. When they kept the men in bed for ten hours, all slept about an hour longer than usual. The additional hour improved their reaction time as well as the attention and vigilance skills needed for driving and monitoring industrial control panels. The men who had been sleepy became more alert, and those already functioning at a high level became even sharper. "Ideally, one would sleep until slept out," Roehrs noted. One then would awaken spontaneously.

HOW MUCH SLEEP IS NECESSARY?

An old saying frowns on sleeping long hours: "Nature requires five, custom takes seven, laziness nine, and wickedness eleven." Physicians used to tell patients, "Everyone needs eight hours of sleep." But studies such as those described above show the fallacy of this generalization.

An individual's sleep requirement is an inherited trait, like height and intelligence; these are not changeable. Short-sleeping mice bred in the laboratory have short-sleeping offspring; the same is true for long-sleeping mice.

Some short sleepers do fine with a minimal amount of sleep that would leave most people reeling. In his Pulitzer Prize–winning book *A Bright Shining Lie*, Neil Sheehan writes of Vietnam war leader John Paul Vann: "His constitution was extraordinary. It permitted him to turn each day into two days for an ordinary man. He required only four hours of sleep in normal times and could function effectively with two hours of sleep for extended periods. He could, and routinely did, put in two eight-hour working days in every twenty-four and still had half a working day in which to relax and amuse himself."

President Bill Clinton is known to be a short sleeper. An article in the Baltimore *Sun* describing the president's first full day in office noted, ". . . for a man who had less than six hours sleep—he got home from the inaugural balls after 2 A.M. and was awakened at 8 A.M.—he kept up the pace impressively throughout the day."

One of the shortest sleepers on record is a seventy-year-old nurse who claimed she had needed only an hour of sleep a day since childhood. When this individual was studied in the sleep laboratory of Ray Meddis at London University, London, England, she averaged sixty-seven minutes of sleep per night for five successive nights. She did not nap and felt alert the rest of the day.

Long sleepers may brag less about their sleep habits than short sleepers and receive less adulation, but they manage just fine, too. Given the pace of modern society, many of these people structure their lives to get nine hours of sleep during the school or workweek and twelve to fourteen hours on days off. Some suffer from their own misguided attempts, or with the help of a well-meaning physician, to reduce sleep time by taking stimulants. Albert Einstein allegedly was a long sleeper. Writer Ann Beattie has said that she likes to sleep ten hours a day.

Short sleepers typically sleep at approximately the same time each day, and get the same amount on both workdays and days off. Adults up to age sixty who regularly sleep less than five hours are classified as short sleepers, and those who get more than ten hours as long sleepers.

The majority of adults may need at least seven hours of sleep per night to avoid the consequences of sleep deprivation, according to the National Commission on Sleep Disorders Research. But in 1983, the most recent year for which statistics were available when *Bodyrhythms* was being written, one out of four Americans aged twenty and above reported getting six or fewer hours of sleep per night. Just six years earlier, in 1977, only one in eight reported getting such a small amount of sleep. Over the past century, Americans have reduced their average nightly total sleep time by more than 20 percent, the commission found. However, there is no evidence, it noted, that people today need less sleep, or that their ancestors slept too long.

Income and other life circumstances may make a difference in the amount of time one is able to sleep. The poor sleep less than the rich, perhaps a reflection of crowded and noisy surroundings. According to the National Center for Health Statistics, about one in three people with annual family incomes of under $7,000 reported getting six hours' sleep or less in 1983. Only one in five people whose family incomes were $40,000 a year or more reported short sleep periods. Additionally, one in three adults having fewer than twelve years of education slept six hours or less, compared with only one in five who had more than a high school education. After age sixty, many people sleep less than they did when younger. Contrary to popular belief, older people do not need less sleep, but they often find it harder to sustain.

Work schedules influence the amount of sleep that people get. According to Michigan State University economists Jeff Biddle and Daniel Hamermesh, working people sleep about an hour less on workdays than those who are not employed. For each additional hour worked, people sleep about ten minutes less. Working women average about thirty minutes less than working men.

Different countries vary in the amount of sleep their citizens get. This difference may reflect the pace of life or the regard with which sleep is held in a particular country. Residents of the United States, as one example, sleep about thirty minutes less a night than those of Belgium.

SLEEP OVER THE LIFE CYCLE

Across the human life span, both the amount and timing of sleep vary considerably. In the first days of life, infants average sixteen to eighteen hours of sleep per day. Some healthy babies, however, sleep as many as twenty-one hours, and others, as few as eleven hours. By six months of age, babies average about twelve hours of sleep per day. Three-year-olds, too, average close to twelve hours of sleep; four-year-olds, eleven hours; and five- to twelve-year-olds, ten hours.

Children aged six to twelve awaken at about the same time each morning and go to bed at about the same time each night. They exhibit higher daytime alertness and more solid sleep than people of any other age. If there were a Sleep Olympics, specialists have said, children aged six to twelve would carry off all the prizes.

During adolescence, the time of life beyond infancy with the most rapid body growth and development, the need for sleep actually increases by about an hour a day, Mary Carskadon of the E. P. Bradley Hospital in East Providence, Rhode Island, has found. But the impact of homework, jobs, and social pressures is so great that teenagers typically sleep an hour or two less per day than preteens.

At a unique summer sleep camp, Carskadon and her colleagues at Stanford University evaluated twenty-seven youngsters from the beginning to the end of their adolescence. When the teenagers were given the opportunity to sleep as long as they could, they averaged about nine hours of sleep per night throughout adolescence. Even with this amount, which was more than they usually got at home, the youngsters became sleepier during the day as they matured. While parents often fret because their teenagers sleep late on weekends, "such behavior is not only normal and appropriate but essential for the majority," Carskadon has observed. "Most teens are not lazy," she said. "They are sleep-deprived."

Carskadon also has found that the changes in biologic clocks that occur during puberty may cause teenagers to go to bed later, and to sleep later, than younger children. She and her colleagues studied nearly five hundred sixth-grade students. Children in this grade are mostly eleven to twelve years old—and are at the same level of social development—but they are often at different stages of puberty. The researchers assessed the children's physical development and the preferred time of day for recess, tests, exercising, going to bed, and getting up. They also noted the children's birth order and other social factors that might have influenced their bedtimes, such as attending a school with older students whose fondness for late-night television shows might have "rubbed off" on the younger students.

The researchers found that the more physically mature students chose to go to bed later on both weekdays and weekends than those who were just starting puberty. While all of the students had to get up at about the same time on weekdays, those who were more physically mature slept later on weekends, the most flexible period in their schedule. These findings counter the prevailing belief that changes in sleep habits during adolescence stem from psychosocial factors, such as a desire to escape parental control.

Moreover, this discovery has important implications for the timing of the school day, suggesting that school should open later for older children than for younger ones. According to Carskadon, the widespread practice in the United States for classes to start earlier in high schools than in junior high schools, and earlier in junior high schools than in primary schools, may run precisely counter to children's biologic needs.

Such findings have worrisome implications for classroom performance. Indeed, when Richard Allen and Jerome Mirabile of Johns Hopkins University surveyed suburban high school students, they found that more than half felt least alert at 10 A.M., and most alert after 3 P.M. They thus attended school during their foggiest hours, and were dismissed from classes just as they reached their peak alertness. Allen also found that high school seniors whose classes started early received lower grades on average than seniors whose classes started later.

Bedtimes are pushed back steadily throughout the high school years, with two thirds of juniors and seniors reporting that they enjoy staying up at night. College students on average go to bed and get up three hours later than high school students. Carskadon's study of the Brown University Class of 1992 from admission to graduation has suggested, however, that this substantial shift in the timing of sleep is a psychosocial one, perhaps reflecting freedom from parental supervision rather than a further change in the schedule of the biologic clock.

The amount of sleep a person needs remains about the same between ages twenty and seventy, but the free-running day shrinks from about twenty-five hours in young adulthood to about twenty-four hours in people in their fifties, sixties, and older. This change imposes an increasingly strong mandate to be "early to bed and early to rise." For some elderly people, this shift in sleep time becomes so extreme that they retire for the night soon after dinner, and awaken at 1 or 2 A.M. Such behavior understandably interferes with social life and gives rise to the frustration of being awake when others are asleep. Fortunately, sleep specialists have devised strategies for dealing with this problem, described in Chapter 6.

THE NATURE OF SLEEP

New understanding of how much sleep people need follows hard on the heels of recent major discoveries about sleep itself. It was at the University of Chicago in 1953 that Eugene Aserinsky, a student working under physiologist Nathaniel Kleitman, found that the eyes sometimes move beneath closed lids during sleep much as they do during waking hours and often faster. Kleitman enlisted William Dement, then a medical student, to pursue this curious phenomenon.

When Dement and Kleitman awakened sleepers at irregular intervals, they discovered that awakenings during episodes of rapid eye movements usually yielded reports of visual, storylike dreams, while awakenings when the eyes were at rest usually did not. This finding dispelled the belief that sleep is one-dimensional. The researchers dubbed the episodes of rapid eye movements *REM sleep*, and soon found that REM sleep alternated about every ninety minutes through the night with periods in which eyes did not move, a state called *non-REM*, or *NREM*, sleep. Studies of the electrical activity of the brain showed that the brain was as active during REM sleep as during wakefulness, sometimes more so. Hence, REM sleep became known as "active sleep," and NREM sleep as "quiet sleep."

While many people still think of waking and sleeping as opposites, the relationship is more akin to adjusting the volume than turning the brain from on to off. Sleep is an active process; it is a state the brain seeks and works hard to maintain. It is not merely the absence of waking, and it is not a time of unconsciousness when the real business of daily life is placed on hold. The simple fact that people generally do not fall off the bed while they are sleeping shows that they stay in touch with their surroundings; sleep consciousness permits one to sleep through the noise of a passing fire siren, even a thunderstorm, but to be roused by the whimper of one's baby. Vivid dreams demonstrate that people are creatures of thought and imagination, both awake and asleep.

A NORMAL NIGHT'S SLEEP

The circadian clock controls when people go to sleep, how long they sleep, and the course of sleep during the night. A night's sleep for most adults usually follows the pattern below. (Also see chart, "A Normal Night's Sleep," on page 69.)

11 P.M.–midnight: Fatigue sets in. Although most people think that they consciously decide to get ready for sleep, their decision actually follows an internally programmed surge in sleepiness. Sleep is so crucial to survival that the body cannot afford to count on its owner to remember to go to bed; as a result, the body sends the message that "now is the time." Sleepiness is a biologic trick, a signal that is hard to ignore. The sequence is not: "I am sleepy; therefore, I ought to go to bed," but "I ought to go to bed; therefore, I am sleepy." The act of going to bed at approximately the same time each day reinforces the likelihood of sleeping at the same time the next day.

Flexibility in one's bedtime may upset the circadian apple cart on occasion, but it also has certain advantages. "Suppose we were 'struck down' by sleep after sixteen or seventeen hours of wakefulness while we were walking home from a late show," Wilse Webb has speculated. "We would have to organize 'sleep patrols' to come and gather up all those who had missed their time schedules from the sidewalks and benches and nooks and crannies where they had dropped into sleep and to deliver them to their beds. We would be trapped in the inexorable demands of a rigid set of boundaries in our life and much of its richness and resources would be lost."

Midnight: Bedtime. The typical twenty- to forty-year-old drifts into sleep within only seven minutes, an astonishingly short time considering the distinct differences between wakefulness and sleep. During the sleep onset period, thoughts wander. One may experience brief, bizarre hallucinations, sometimes called *dreamlets* or *hypnogogic reverie*, and have the sensation of falling or floating. Some muscles relax sooner than others, and this unevenness may trigger a sudden jerk of an arm, leg, or even the entire body.

12:07–12:30 A.M.: NREM episode 1. Sleep ordinarily begins with NREM sleep. For the first fifteen minutes or so, known as NREM Stage 1, most people will respond to the query "Are you asleep?" with a denial. They will insist, "I was just lying here thinking." The pattern of brain waves recorded at the start of sleep resembles small, closely spaced peaks and valleys.

12:30–1 A.M.: Brain waves become farther apart, larger, and more regular. Bursts of high-voltage brain activity that last a second or two, called *sleep spindles*, mark the border between Stage 1 and Stage 2. As sleep deepens, heart rate

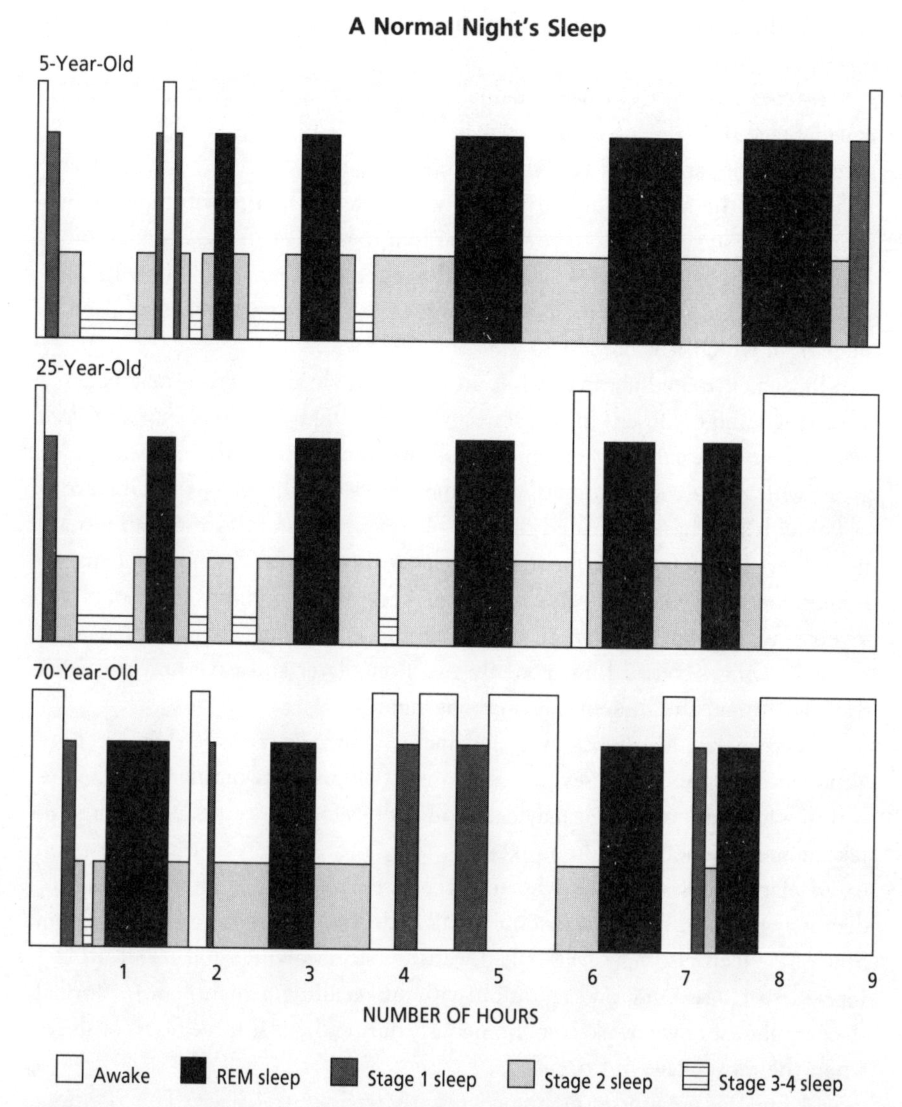

Healthy young adults spend about 10 percent of their sleep time in Stage 1, 50 percent in Stage 2, 20 percent in Stage 3-4, and 20 percent in REM sleep. Children may spend as much as half of their time in REM sleep, a finding that suggests that this stage, with its high mental activity, is necessary for learning and memory. As people grow older, the proportion of time spent in NREM Stage 3-4 dwindles, and that in Stage 1 and Stage 2 increases. By age sixty, most people spend only about 1 percent of their sleep time in Stage 3-4, a fact that may account for the frequent complaint that sleep becomes less restful with age. The proportion of time in REM sleep, however, remains relatively stable from late childhood until people are in their eighties, when it declines slightly and is replaced by NREM Stages 1 and 2.

and breathing slow down, blood pressure falls, and temperature continues its downward path.

1–1:45 A.M.: NREM sleep gradually progresses to its deepest level. This is called Stage 3-4, the name a carryover from the early days of sleep research when scientists attempted to distinguish Stage 3 from Stage 4 by the size and frequency of the slow, undulating brain waves etched by the polygraph. However, the two stages proved to be so similar that today they are usually combined. People deprived of Stage 3-4 sleep will be extremely sleepy the following day, which has led to the hypothesis that this stage is the most restful. It is also the stage from which it is hardest to awaken.

Indeed, if roused abruptly while in Stage 3-4 sleep, a person may become disoriented and confused. In this state, which is called *sleep drunkenness* or *sleep inertia*, he or she may not be able to talk or think coherently, or even move about with the usual coordination. Studies by David Dinges of the University of Pennsylvania have indicated that people may perform far worse in this state than they did after going for many hours without sleep. In one of Dinges's studies, a college student who had been kept awake for fifty-two hours got seventy-five correct answers on a three-minute subtraction test. He then went to sleep. Dinges roused him suddenly two hours later and gave him the same test. This time, the student got only one right.

Sleep inertia also takes away judgment; people may insist they are fine, although they are in fact severely impaired. Sleep inertia ordinarily dissipates within a few minutes, but it may cause difficulties for those who are required to take immediate action upon awakening, such as firefighters who must respond to an alarm, military pilots who must be airborne minutes after an alert, or doctors who must make an instant medical decision. Partial awakenings from Stage 3-4 sleep also may trigger bed-wetting, sleepwalking, and terror attacks (episodes of screaming and agitation without recalling a nightmare). Normal sleepers almost never awaken spontaneously during the first three hours of sleep, when the most Stage 3-4 occurs.

A burst of growth hormone occurs early in Stage 3-4 sleep. This hormone is necessary for normal childhood development, and it aids in the healing of wounds and in other bodily repairs in adults. The biologic clock for the release of growth hormone is tied to the one that controls sleep. If sleep is fragmented—as occurs, for example, in some victims of child abuse—the release of growth hormone is disrupted; indeed, children deprived of sleep may not grow normally. When removed from the abusive situation, they may experience dramatic growth spurts.

1:45–1:55 A.M.: REM episode 1. About ninety minutes after the start of sleep, most adults will rise from deep NREM Stage 3-4 into the light sleep of Stage 2 for a few minutes, then roll over and enter their first REM episode, in which they will experience the first of the four or five dreams of the night. REM sleep is so different from NREM sleep that some sleep specialists describe it as a third state of existence. Its brain waves, fast and irregular, resemble those of wakefulness.

While people may shift positions during NREM sleep about thirty times a night, they lie virtually paralyzed in REM sleep, perhaps so that they do not act out their dreams. If awakened abruptly from REM sleep, a person may experience a transitory sense of paralysis. People normally awaken for a few seconds at the end of each REM episode, usually too briefly to remember doing so. This may be an evolutionary "leftover." Most animals awaken frequently to check that their surroundings are safe. The final awakening in the morning, if spontaneous, usually takes place right after REM sleep.

About 80 percent of awakenings from REM episodes will yield dream reports, while only about 20 percent of those from NREM episodes will do so. Awakenings from REM sleep produce fanciful tales: "I was strolling through an underwater city"; "I was flying." NREM reports are more often mundane: "I was planning my grocery list"; "I was thinking about some work I have to do."

While everyone has several REM episodes and thus presumably several dreams a night, few dreams are remembered. The dream most likely to be remembered is the last one of the night. The usual explanation for poor dream memory is that it is hard to translate the picture language of the night into the verbal language of the day; Freud posited that dreams dredged up material a person would prefer not to remember and therefore needed to work to recover them.

Some contemporary theorists have suggested that dreams are not worth remembering at all. Nobel Prize–winner Francis Crick and Graeme Mitchison, a British mathematician, have hypothesized that the mind has only limited storage space and that dreaming helps purge it of useless information.

Others, however, believe that dreams help solve problems of daily living. Rosalind Cartwright of the Rush-Presbyterian-St. Luke's Medical Center in Chicago has reported that the dream content of a given night may have its own rhythm, particularly when people are under stress. In a study of men and women in the process of divorcing, she found that the first dream of the night often reflected a recent experience. The next dream returned to a similar event earlier in life. The next two or three integrated the two experiences, jumping ahead to the future and tying up loose ends in the present. This rhythm lost its beat in people who were depressed. Then dreams became shorter and less imaginative, often repetitive, like a phonograph needle stuck in a record groove.

Indeed, the time when REM sleep occurs also changes when people are depressed. David Kupfer of the University of Pittsburgh and his colleagues have found that REM sleep in people with depression begins much earlier than is normal, often soon after the start of sleep. The earlier the REM period, Kupfer has discovered, the more severe the depression is likely to be. One of the newest findings from chronobiology studies is that changes in the REM sleep clock may do more than reflect depression; they actually may trigger it or make it worse. (For more on depression and the biologic clock, see Chapter 7.)

1:55–3:30 A.M.: NREM episode 2. The sleeper travels through Stage 2, returns to Stage 3-4, and the cycle repeats. Temperature regulation, having been cut off temporarily during REM sleep, now restarts. In the first half of a sleep episode, temperature normally follows a downward course.

3:30–3:45 A.M.: REM episode 2. REM episodes grow longer over the night. Sleep specialists sometimes liken the first dream of the night to a preview of coming attractions, the second and third to short subjects, and the fourth and—if sleep continues long enough—the fifth, to feature films.

3:45–5 A.M.: NREM episode 3. This episode will contain more Stage 2 and Stage 1 sleep and less Stage 3-4 than NREM episodes 1 and 2.

4 A.M.: This is midnight on the body clock, the time of lowest body temperature. A person who is awake at this hour experiences the highest degree of sleepiness and thus the poorest performance. This is when telephone operators take longer to answer calls, meter readers jumble more numbers, train drivers miss more warning signals, drivers have more single-vehicle accidents (those most likely to reflect lapses in attention), and doctors misread more electrocardiograph tapes.

This is also the time of greatest sensitivity to the rhythm-setting effects of light. An insomniac who turns on a light to read in the middle of the night runs the risk of priming his or her body clocks for an unwanted wake-up call at the same time the following night, according to studies by Charles Czeisler and his colleagues at Harvard University. When body temperature is low, even ordinary lamplight may fool the body into thinking sunrise has come early.

If one changes one's usual time for sleep, the temperature rhythm will shift too, but it changes so slowly that the predawn hours will remain the sleepiest hours of the day for a week or longer. People who move to Rome and live as the Romans do will adapt more quickly to the change in lifestyle than the typical night- or rotating shift worker in an industrialized society who must stay up at night and sleep during the day but on days off reverts to sleeping during the night. (For more on this topic, see Chapter 10.)

5–5:30 A.M.: REM episode 3.

5:30–6:30 A.M.: NREM episode 4, a time of increasingly lighter sleep.

6:30–7:15 A.M.: REM episode 4, often terminated by a final awakening. In REM sleep, heart rate, breathing rate, and blood pressure show more variability than in NREM sleep. The pulse rate, for example, may soar from a normal resting rate of 70 beats a minute to 140 beats a minute. Such variability could be viewed as a physiologic betrayal, a challenge to the comforting and familiar view that sleep is a time of rest. It is not by chance that death occurs more often in the early-morning hours, when REM sleep is most frequent. (For more on illness and time of day, see Chapter 8.)

In REM sleep, males of all ages experience erections, and females, clitoral engorgement. These changes occur regardless of dream content. Some researchers hypothesize that they represent part of the body's automatic preventive-maintenance program, a physiologic self-test. Men who awaken in the morning with an erection sometimes attribute it to a full bladder, but it is more likely to be a carryover from the night's last REM episode. Specialists in sexual disorders often counsel men with erectile difficulties to take advantage of this phenomenon.

THE NINETY-MINUTE DAY

Recognition of the rhythms of sleep prompted questions about sleepiness and alertness rhythms all around the clock. To study these rhythms, in 1980 at Stanford University, Mary Carskadon and William Dement created what they called "a ninety-minute day." Volunteers went to bed every ninety minutes around the clock, staying in bed for thirty minutes each time. Had they been able to fall asleep quickly and stay asleep, they would have accumulated eight hours of sleep over the course of a twenty-four-hour day. But none did. Sometimes they fell asleep fast, and sometimes not at all.

This study led to the development of the standard measure of sleepiness and alertness that is used in sleep laboratories everywhere. Known as the Multiple Sleep Latency Test, or MSLT, it follows a routine similar to that of the original study but uses a 120-minute cycle. If people manage to fall asleep, they are awakened after ninety seconds to keep them from accumulating enough sleep time to affect later studies. If they do not fall asleep, their nap opportunity terminates after twenty minutes. The MSLT shows that it is easy for most people to fall asleep at some times, such as midafternoon, and hard at others, such as midmorning and, surprisingly, midevening, just two or three hours before bedtime.

MSLT studies also established that sleepiness is cumulative. That is, people who shortchange themselves even modestly on sleep on weeknights may become seriously sleepy by the end of the week. Factor in the synergistic effect of sleepiness and alcohol, and it is easy to understand why alcohol-related traffic accidents peak on Friday night. One might expect that such accidents would be more common on Saturday night since more people drink then. However, many people are actually more alert on Saturday night than on Friday night because they sleep later on Saturday morning. (And because they also sleep later on Sunday morning, noontime Sunday is the safest time during the whole week to drive.)

While taking the MSLT, most people got a few minutes of sleep. Did that little bit alter alertness? To find out, Peretz Lavie and his colleagues at the Technion-Israel Institute of Technology shrank the experimental day to only twenty minutes. Their studies illuminate minute details of the daily ups and downs.

In their studies, volunteers went to bed every twenty minutes, but stayed in bed only seven minutes. After rising, they took a variety of mood and performance tests. The volunteers were tested when they had just awakened from a normal night's sleep, at the end of their normal working day, and after they had missed part or all of the night's sleep. Some were asked to get into bed and try to fall asleep, and others, to get into bed and try to resist sleep. The responses to all these different maneuvers consistently showed that people did not become increasingly sleepy as the day wore on. Rather, it was a stepwise process; the subjects became more sleepy, then less sleepy, then more sleepy, and so on.

Over the day, there was a series of what Lavie called *sleep gates*, approximately 90 to 120 minutes apart. When these gates to sleep were open, people fell asleep easily. When they were closed, people seldom fell asleep.

To illustrate, for most people the gates are closed approximately one to three hours before their usual bedtime; indeed, this period is virtually a forbidden zone for sleep. Someone who normally goes to bed at midnight has a forbidden zone from 9 to 11 P.M. Staying up until 2 A.M. for a few nights would shift the forbidden zone two hours later. Upon returning to a midnight bedtime, that person might have trouble falling asleep for several nights until a realignment occurs.

Even during the night, Lavie found, people fell asleep faster at some times than at others. That is why even a modest change in the time one attempts to go to sleep may make sleep more restless than usual. (For more on insomnia, see Chapter 6.)

THE VALUE OF NAPS

The ultrashort sleep/wake schedules also demonstrate that the desire for a midday nap is perfectly normal. "We are programmed to have a long sleep period at night, and a short one in the middle of the afternoon," according to Roger Broughton of the University of Ottawa. This programming suggests that naps are functional.

Famous nappers include Wolfgang Amadeus Mozart, Napoleon Bonaparte, and John D. Rockefeller. "I always went to bed at least for an hour as early as possible in the afternoon and exploited to the full my happy gift of falling almost immediately into deep sleep," Winston Churchill said. "By this means," he claimed, "I was able to press a day and a half's work into one."

Napping is common in tropical climates and in farm and rural settings. Although the word "siesta" means the sixth hour, counting from sunrise, naps are customarily taken in midafternoon. However, even given the opportunity to nap, as well as a society that encourages it, not everyone does. Studies in Italy, Greece, and Mexico show that only about two out of five people in these traditional siesta cultures nap four or more times a week. That number is falling, researchers say, as cities grow larger and commuting takes up a larger part of the day. Capitalism has instilled the idea that time is money; this has dampened interest in taking time for non–income-generating activities, such as naps.

The study of people who can set their own agendas offers insight into the sleep schedules that more people might prefer if they could organize their time as they wish. More than half of all college students, for example, nap one or more times a week.

David Dinges found that college students nap for different reasons. About three out of four said that they napped only to make up for sleep lost at night. These Dinges described as "Compensatory," or "C-nappers." The others napped simply because they enjoyed it. These he dubbed "Recreational," or "R-nappers."

The C-nappers snoozed one or two afternoons a week. The R-nappers napped more frequently, about three to five times a week; they were even more apt to do so when they missed some sleep at night. Both groups reported satisfactory nighttime sleep. Long sleepers napped more often than short sleepers, most likely a reflection of an overall greater sleep need. According to psychological profiles, both groups were equally productive, active, and energetic. The R-group became no more bored or depressed than the C-group. They did not, as one might imagine, use naps to escape from their problems.

Among habitual nappers, the average nap lasts seventy to eighty minutes.

Salvador Dalí claimed to be refreshed by the briefest of catnaps—the time that it took a spoon, held in his hand while he fell asleep, to drop onto a tin plate on the floor, where it would clang and wake him up. Sleep laboratory studies show, however, that a nap must last at least ten minutes to affect mood and performance.

Naps are not miniatures of a normal night's sleep, which starts with mostly deep sleep and becomes lighter as the night progresses. Rather, the naps one takes early in the day resemble the tail end of the nighttime sleep period, with light sleep dominating. Naps taken late in the day resemble the start of the nighttime sleep period, with deep sleep dominating. Because people descend gradually into sleep, brief naps contain little of sleep's deeper stages. Napping early in the day, when alertness is already high, will not improve alertness. Napping late in the day may induce grogginess and make it harder to fall asleep at night. Napping in midday results in the greatest enhancement in alertness. For someone who has missed some sleep, however, a nap at any time of day will improve alertness; even a little sleep is better than no sleep.

Perhaps the most astonishing—and practical—finding to emerge from nap studies is that sleep, like money, can be put in the bank. People who expect to suffer a midweek sleep loss—as a result of having to complete a work assignment, say—will function better after having slept longer the weekend before, Dinges has demonstrated. People who anticipate a late night will perform better that night if they take a midday nap than if they do not. These findings have direct application to the workplace.

One rather startling study (imagine the steps it must have taken to secure approval to proceed!) found that pilots who napped in the cockpit for forty minutes during the cruise portion of transoceanic flights were more alert for the remainder of the flight than nonnapping pilots. Compared to the latter group, the pilots who had rested scored higher on a variety of tests of vigilance and attention, according to researchers for the National Aeronautics and Space Administration and the Federal Aviation Administration. Brain-wave studies, conducted with portable monitors, also showed that the pilots who napped were more alert later during the flight. This suggests that people may derive as much benefit from a nap break as from a coffee break. Perhaps even more.

CHAPTER 5

THE RHYTHMS OF SEXUALITY

THE CLOCK THAT GOVERNS ALERTNESS and sleepiness also regulates sexual rhythms in both sexes, from hormone release to desire and performance, each day, each month, each year, and throughout a person's lifetime.

SEXUAL RHYTHMS IN WOMEN

The most familiar of all sexual rhythms is the menstrual cycle—the monthly production of eggs by the ovaries, preparation of the uterus for pregnancy, shedding of built-up tissue and some blood through the vagina if pregnancy does not occur, and start-up of the process again.

The words "menstruation," "moon," and "month" all stem from the Greek word for "measurer of time." The relationship among the three is not merely etymological; like circadian rhythms, monthly rhythms in the body mirror those of the natural world. A woman's menstrual cycle is about the same length as the lunar cycle; moreover, periods are more likely to start during the week of a full moon than at any other time of the month. Indeed, moonlight may help synchronize the menstrual cycle and, as described on pages 83–85, chronobiologists have found that artificial moonlight may help normalize irregular cycles.

Menstruation marks a woman's fertile years. The start of menstruation, called *menarche*, generally takes place between ages ten and sixteen, or whenever a girl acquires a certain amount of body fat and weighs about one hundred pounds. When the biologic clock for reproduction runs down, the end of menstruation, called *menopause*, occurs, usually between ages forty-five and fifty-five. During her lifetime, the average woman thus will have about four hundred cycles, with about six days during each one when conception is possible. She will experience bleeding each cycle, except during pregnancy, breast-feeding, and some illnesses.

Sexual functioning in both sexes is controlled by a hormonal chain reaction. As hormones are secreted into the bloodstream, they form a feedback loop. The key sexual hormones in women are estrogen and progesterone, both produced by the ovaries. The comparable hormone in men is testosterone, produced mainly by the testes. Testosterone is also produced by the adrenal glands, located just above the kidneys. Hence, both men and women manufacture testosterone, although men make ten times as much; a high testosterone level for a woman would be a low rate for a man. In women, testosterone stimulates bone growth at puberty and prompts growth of underarm and pubic hair. It also helps foster sexual desire throughout a woman's lifetime.

The secretion of these hormones is directed by follicle stimulating hormone-releasing factor (FSH-RF) and luteinizing hormone-releasing factor (LH-RF), which are both produced by the hypothalamus; and by follicle stimulating hormone (FSH) and luteinizing hormone (LH), which are both produced by the pituitary gland, which is attached to the hypothalamus by a stalk.

All of these hormones have daily patterns of release; LH, for example, is secreted in larger quantities at night. Because the hypothalamus is the home of the body's master clock, the SCN, events that cause disturbances in the SCN, such as sleep-cycle changes, jet travel, and shift work, may affect sexual rhythms as well.

A TYPICAL MENSTRUAL CYCLE

In the first year or two of menstruation, the production of eggs is somewhat erratic, and cycle length is often irregular. Once the number of days per cycle becomes stable, the cycle usually lasts about twenty-eight days, although a cycle averaging a week more or less is perfectly normal. For any one woman, the number of days per cycle is relatively consistent. As menopause nears, the production of eggs slows down, and the cycle typically lengthens and becomes more irregular. When the ovaries stop producing eggs, bleeding stops.

A typical twenty-eight-day cycle follows the schedule below:

Days 1–5: Menstruation. The first day of the cycle is the day when bleeding begins. In the absence of a fertilized egg, the lining of the uterus breaks down and is shed as the menstrual flow. On the first day of bleeding, estrogen and progesterone are at their lowest levels during the cycle. The low levels of these hormones prompt the hypothalamus to step up production of FSH-RH, which in turn stimulates production of FSH. FSH tells egg-containing pockets or follicles of one ovary—rarely, both—to prepare several eggs for release. The follicles start to produce estrogen, which signals the uterus to thicken in preparation for the arrival of an egg. The part of the cycle from the onset of menstruation until the egg is released is called the *follicular phase*.

Days 6–14: Estrogen levels continue to rise; the walls of the uterus continue to thicken. Progesterone remains low until ovulation, or the release of an egg—occasionally, two or more eggs. The time needed for eggs to mature may vary by several days from woman to woman and from cycle to cycle.

The uncertainty about when ovulation occurs decreases the reliability of the rhythm method of birth control, which is based on avoiding unprotected sexual intercourse for at least three days before ovulation and three days afterward—the amount of time that sperm can live in the female reproductive tract. The rhythm method works best in women whose menstrual cycles are extremely regular, and even for them, the variability of egg maturation makes it less effective than many other forms of birth control, including condoms, and oral and implanted contraceptives.

Day 14: Estrogen reaches a high level, prompting the hypothalamus to produce a burst of LH-RF. This hormone triggers the release of LH, which stimulates ovulation. Ovulation may be accompanied by a cramplike pain. The ruptured follicle will form the corpus luteum, a factory for the manufacture of progesterone. Egg-containing follicles that did not release an egg will shut down and shrink. The part of the cycle from ovulation to the start of menstruation is known as the *luteal phase*; it runs a relatively fixed fourteen days.

Estrogen is at its highest level right after ovulation. High levels of estrogen have been linked to several positive effects on both physical and mental functions. The senses of vision and smell, for example, are most acute at ovulation; women are able to detect faint lights and certain odors better than at other times of the month. They are least bothered by pain. Feelings of well-being and self-esteem are also highest. Many women report their most intense feelings of sexual desire at this time. (For more on sexual desire, see page 87.)

Ovulation has a noticeable impact on the content of dreams: before ovulation, when impregnation is possible and estrogen levels are rising, active dreams predominate, winning a prize, for instance; after ovulation, assuming the possibility of impregnation has passed and estrogen levels are falling, dreamers are more likely to be passive than active, dreaming, perhaps, of being chased by a potential attacker. Dreams just before menstruation, when estrogen is at its lowest level, score highest in anxiety and depression. Dreams during menstruation—a time of abstinence from sexual activity for many women—are highest in explicit sexual content.

Days 15–18: The mature egg travels down the fallopian tubes; where fertilization, if it is to occur, will take place.

Days 15–22: Meanwhile, LH has prompted the ruptured ovarian follicle to manufacture progesterone. This hormone turns the lining of the uterus into spongy, blood vessel–rich tissue that would be hospitable to a fertilized egg.

The appearance of progesterone in the bloodstream is accompanied by a rise of about 0.5°F in basal body temperature. An elevation in temperature for three days in a row signals that ovulation most likely has occurred, and that unprotected intercourse after this time will probably not result in pregnancy. Women who wish to detect this temperature elevation often are advised to take their temperature each morning after at least five hours of sleep while still lying in bed.

This method, however, is not entirely reliable. Day-to-day temperature variability may occur even in women who scrupulously set their alarms for the same time, and do not move around before taking readings, and even when readings are taken in hospitals. Chronobiology research has provided an explanation—and a solution: the early-morning hours are those most dense with REM sleep, a state when temperature may vary considerably. According to Kathryn Lee of the University of California at San Francisco, readings taken soon after REM sleep may reflect that variability. Her studies suggest that for most women, readings taken just prior to sleep onset may be a more accurate indicator of ovulation. (For instructions on taking readings, see "Self-Test: Alertness and Temperature Cycles," page 47.)

Days 23–27: By this time, the egg either will be implanted, and pregnancy will proceed, keeping levels of both progesterone and estrogen high, or it will not, which causes both hormone levels to drop. Temperature remains elevated as long as progesterone levels remain high. When the levels of both hormones become sufficiently low, the thickened lining of the uterus will break down. The unfertilized egg will disintegrate or pass from the body unnoticed.

During this time, many women experience physical and emotional symptoms of what is generally called *premenstrual syndrome*, or PMS. Common symptoms include breast tenderness, weight gain, and irritability, but the list is far-reaching. "Nearly every symptom ever attributed to any illness has been reported to be a symptom of PMS," according to Sally Severino and Margaret Moline of Cornell University. In reviewing some nine hundred publications on the topic for their book *Premenstrual Syndrome*, Severino and Moline cataloged more than one hundred symptoms.

Some women report that they experience positive symptoms in the week or so before menstruation. These include increased introspectiveness and sensitivity, greater creativity, and increased interest in sex. Writer Joyce Mills has called the week before menstruation a time of PMA—premenstrual awareness.

But the majority of women report negative symptoms. Many of these are attributed to progesterone, which is at its monthly peak at about day 22; the amount of this hormone that is released is about twenty-five times higher at this time than it will be just after menstruation starts. By causing swelling in the glandular tissue of the breasts and congestion in breast blood vessels, progesterone may produce tenderness in the breasts, weight gain, and fluid retention. As an appetite stimulant, progesterone may account for premenstrual food cravings, particularly for carbohydrates. Animal studies show that progesterone has a sleep-inducing effect, perhaps a factor in the often-reported need for extra sleep and increased daytime sleepiness. By raising body temperature, however, progesterone may disrupt sleep, inducing irritability, mood swings, and depression.

The constellation of emotional symptoms prompted the American Psychiatric Association (APA) to list severe PMS as a depressive disorder in its 1994 update of the psychiatrists' bible, the *Diagnostic and Statistical Manual of Mental Disorders*. (The APA calls it *premenstrual dysphoric disorder*; see symptom checklist, page 82.) Inclusion of this diagnosis was much debated for many years within the profession and without, since labeling as a "disorder" symptoms that may affect half the population colors the way scientists and others report and interpret research.

"With women comprising a substantial and increasing proportion of the work force," Severino and Moline observed in their book, "medical problems that could have a bearing on performance and productivity or that could be used to discriminate against job placements or promotions for women have become 'hot potatoes.' " Nonetheless, relatively few women—probably less than 5 percent, according to the APA and the American College of Obstetricians and Gynecologists—report that premenstrual symptoms interfere with their work, relationships, or normal activities.

Kathryn Lee has found that women who experience premenstrual mood changes spend less time throughout their menstrual cycle in the deeper sleep of Stage 3-4 and more in the lighter sleep of Stages 1 and 2 than women of the same age who do not experience premenstrual mood changes. For this reason they may be more vulnerable to the disruptive effects of progesterone on sleep when it reaches a high level in the bloodstream.

The balance of estrogen to progesterone also may play a role in causing

PREMENSTRUAL DYSPHORIC DISORDER

Symptom Checklist

To be diagnosed as having premenstrual dysphoric disorder,* a woman must have experienced at least five of the following symptoms in most menstrual cycles during the past year. The symptoms must have occurred most of the week before menstruation began and markedly disrupted work, school, usual social activities and relationships.

Also, the symptoms must have started to abate within a few days of the onset of menstruation, and they must have been absent in the week following a period. At least one of the symptoms must have been 1 through 4 below:

1. Markedly depressed mood, feelings of hopelessness, or self-deprecating thoughts
2. Marked anxiety, tension
3. Marked lability in moods (e.g., suddenly feeling sad or tearful)
4. Persistent and marked anger or irritability or increased interpersonal conflicts
5. Decreased interest in usual activities
6. Difficulty concentrating
7. Easy fatiguability or marked lack of energy
8. Marked change in appetite, overeating, or specific food cravings
9. Excessive sleep or insomnia
10. Feeling overwhelmed or out of control
11. Other physical symptoms, such as breast tenderness or swelling, headaches, joint or muscle pain, "bloating," weight gain

*Premenstrual dysphoric disorder is listed among diagnostic categories needing further study in the 1994 edition of the American Psychiatric Association's *Diagnostic and Statistical Manual of Mental Disorders*. Mental health professionals must use diagnoses listed in the manual to support claims for medical insurance reimbursement.

Adapted with permission from *Diagnostic and Statistical Manual of Mental Disorders, Fourth Edition*. Washington, D.C.: American Psychiatric Association, 1994, p. 717.

PMS, as suggested by the finding that women who take birth control pills report they have fewer PMS symptoms than women not taking such pills; the steady dose of estrogen and progesterone that those who take those pills receive may confer some protection. Replacement therapy using various combinations of estrogen and progesterone has not proven a remedy for PMS, however, which suggests that these hormones are not the only culprits in the disorder.

Women with certain chronic illnesses, including asthma, epilepsy, and migraine headaches, may find that their symptoms worsen premenstrually. The presence—or absence—of estrogen at different times of the month may influence the success of surgery for breast cancer. (For more on the relationship of the menstrual cycle to various illnesses, see Chapter 7.)

Some researchers have suggested that the menstrual cycle itself is a zeitgeber that triggers PMS. According to this theory, the expectation that PMS symptoms will appear creates a self-fulfilling prophecy. The shifts in hormonal activity that occur at the same time may be merely coincidental. In one study, researchers took blood samples from a group of women, telling some that the tests showed that their periods would start in a day or two. In fact, these women had just ovulated and would not menstruate for fourteen days. Some of them developed PMS symptoms anyway.

Day 28: The uterus sheds its now unneeded lining along with some blood. Uterine contractions expel the lining through the cervix into the vagina. Menstruation begins, and the cycle restarts.

LIGHT AS A MENSTRUAL ZEITGEBER

Light may play an important role in synchronizing the menstrual cycle. Put another way, menstrual irregularity, even infertility, may be the result of extreme variations in or low levels of light exposure. Recent findings about light and fertility have a "Believe It or Not" flavor, yet they are too consistent to be ignored. In industrialized societies, where ever-larger percentages of the population spend most of their time indoors, fertility rates have been falling. Frequency of fertility is lower in people who are blind than in those who are normally sighted. Fertility is also lower in women who live in extremely northern latitudes, such as Finland, in the so-called dark months; and it rises in those regions in the months of midnight sun.

These findings may be tied to the still-mysterious hormone melatonin, which is produced by the pineal, a pea-sized gland at the base of the brain. In some birds and reptiles, the pineal is so close to the top of the skull that it detects light directly, and for this reason it has been called "the third eye." In mammals, including humans, the pineal is buried too deeply to receive direct light stimulation. Instead, light signals reach the pineal by traveling from the eyes over nerve pathways. Darkness prompts the pineal to secrete melatonin; light causes it to stop secreting the hormone.

The length of melatonin secretion thus varies with the seasons, becoming longer in the fall and shrinking in the spring. In some animals, these changes serve as the on/off switch for mating behavior; depending on the species, the same signal may have the opposite effect. As days grow darker in the fall, sheep start breeding, for instance, while Syrian hamsters stop. Melatonin may figure in human fertility cycles, too.

Light exposure may prompt menstruation to start at the appropriate time. Girls who were blind at birth, or became blind in their first year of life, start menstruating sooner than is average for those with normal vision. Women living at sea level (where light intensity is low) start menstruating sooner than those who live at higher altitudes.

Even more extraordinary, and potentially enormously useful, is the discovery that exposure to even a small amount of light, that from a single hundred-watt bulb—the equivalent of light from a full moon—may normalize irregular periods. In 1967, Edmond Dewan treated a woman who had had irregular periods for sixteen years. He advised her to leave a lamp on all night in her bedroom on nights 14 through 16, counting from the day her last period began. The lamp was positioned so that a single hundred-watt bulb would reflect light off the walls and ceiling onto her face throughout the four-month study. Although her cycles had previously varied from three to nine weeks, they soon assumed a regular twenty-nine-day rhythm.

This study has important implications for both women who are trying to conceive and those who are trying to avoid conception, but no one took it seriously until chronobiologists saw those possibilities. In 1990, May Lin and her colleagues at the University of California at San Diego (UCSD) repeated and expanded Dewan's experiment, using sixteen volunteers with long or irregular menstrual cycles.

They asked seven volunteers to place a lamp with a hundred-watt bulb three feet from the head of their beds and to leave the light on all night on days 13 through 17 of their menstrual cycle. Nine others used a dim red photographic darkroom bulb in the same way. The women using the hundred-watt bulb read in bed for thirty minutes before going to sleep to ensure some waking light exposure; the red light was too dim to permit reading. The women using the hundred-watt bulb developed shorter and more consistent menstrual cycles, which fell from about forty-six days to thirty-three days. After they stopped using the light, however, their cycles lengthened again. Those who were exposed to the dim red light showed no changes in their cycles.

In a second, similar study in which the lights were used on days 10 to 14, a team of UCSD researchers, led by Michael Drennan, found that the benefits of the hundred-watt light were the same as in the earlier study, and moreover, that the women's cycles did not lengthen appreciably afterward. This suggested the possibility that there was a carry-over effect when the light exposure came a few days before, as well as on the day of, ovulation. Light had no effect on a group of women with normal-length cycles participating in the same study.

In a third UCSD study, the effects of bedside lamps were compared with those of an electronic mask containing either an equivalent bright or dim light that turned on gradually about thirty minutes after a woman went to bed, and turned off just before she usually awakened in the morning. The mask, designed by Roger Cole of UCSD, was worn on days 13 through 17. Katharine Rex and her colleagues found that when the lights were placed close to the subjects' eyes, even the dim red light, which is comparable to pale moonlight, normalized the cycle length.

How could light have such a dramatic impact on the length of a woman's periods? The leading theory is that normal cycle length depends on the secretion of normal amounts of melatonin. Women who have irregular periods may produce higher amounts of this hormone than their bodies require. Light exposure may serve to turn off the spigot at the crucial time of the month.

Light also may relieve symptoms of PMS. In their first study, Barbara Parry and her colleagues at UCSD found that some women with PMS symptoms got better after they spent two hours in the evening in front of a bank of bright lights with an intensity akin to sunlight. But in a later study, the researchers found women with PMS improved just as much after exposure to bright light in the morning, and even after exposure to dim red light in the evening. The dim light had been used with the expectation that it would serve as an inactive comparison, or *placebo*. Thus, all of the light treatments could be placebos, or all could be physiologically active, a controversy that will necessitate further research. (For other strategies for self-help for PMS, see the Appendix.)

SOCIAL ZEITGEBERS AND THE MENSTRUAL CYCLE

Women who live or work together often have periods at the same time. This phenomenon is most familiar in women who are related to each other—mothers and daughters, or sisters—but it may also occur in close friends or co-workers. For example, seven female lifeguards had their periods at scattered times when the summer began. After they spent three months together, all found that their periods began within four days of the others'.

In 1971, Martha McClintock of Harvard University charted this phenomenon, which came to be known as the *dormitory effect.* She studied menstrual cycles in 135 women aged 17 to 22, all residents of the same dormitory at a suburban women's college. Between September and April, the women reported the dates that their periods began. They also identified by room number the women with whom they spent the most time. As the year progressed, the start of periods in roommates and close friends moved nearer together. The critical factor was the amount of time the women spent with each other. The geographic location of the women's rooms played no role.

McClintock also asked the women how much time they spent with men each week. Many reported that their cycles became more regular and more frequent when their dating increased. One woman said she had periods only once every six months until she started dating regularly. Her cycle length then shortened to four and one half weeks. When she broke up with her boyfriends, the cycle lengthened again. McClintock interpreted these findings as indicating that "there is some interpersonal physiological process which affects the menstrual cycle." She was not able to clarify what it was about close contact that had such a dramatic effect.

Present-day chronobiologists might point to social zeitgebers as the explanation. The women presumably attended classes with their friends, shared meals with them, participated in social events and studied together, and even slept on the same schedules. All these activities may have served to regularize a panoply of biologic rhythms, including those of the hormones governing the menstrual cycle.

ODORS AS ZEITGEBERS

Additionally, the women's menstrual clocks may have been reset by exposure to certain natural body odors, or pheromones, from their friends. Evidence that pheromones may control the timing of the menstrual cycle comes from a series of studies at the Monell Center of the University of Pennsylvania.

In one study, George Preti and his colleagues collected underarm secretions from a group of women throughout their menstrual cycle. The researchers pooled the secretions taken at the same point in the cycle, mixed them in ethanol, and exposed other women to them by rubbing the extract under their noses three times a week. Within three to four months, the recipients synchronized their cycles to match those of the donors of the secretions. Women who sniffed ethanol alone, presented the same way, did not change their cycles.

Pheromones may be more powerful than social cues in regulating timing. In Preti's study, the odors exerted their influence in the absence of other social contact; indeed, the researchers found that even mixtures from different women and different cycles altered the timing. The crucial factor was the time of the cycle that the secretions were obtained.

Pheromones may also explain why women who engage in regular (at least weekly) sexual activity with men have more regular menstrual cycles than women who do not. Male contact may prime a woman for fertility. In another University of Pennsylvania study, Winnifred Cutler and her colleagues found that even women who were not having regular sexual activity developed more regular menstrual cycles after being exposed to underarm secretions taken from men, prepared and presented in the same way as in the Preti study. The women in the Cutler study had previously had cycles of less than twenty-six days or more than thirty-two days; within three to four months of exposure to the male scents, their cycles averaged about twenty-nine and one half days. Intimate contact with men is apparently necessary to alter menstrual rhythms. All the women in this study spent much of their day in a heterosexual environment, but only those who were exposed to the male underarm extracts developed regular cycles.

RHYTHMS OF SEXUAL ACTIVITY

Studies across the menstrual cycle show that sexual intercourse is least frequent during menstruation and most frequent around day 8. This rhythm may reflect both social and biologic factors: On the social side, many couples abstain from intercourse during a woman's period; the cessation of bleeding thus prompts a return to sexual activity. The biologic impetus may stem from rising levels of estrogen, a hormone that makes females more likely to initiate sex.

After day 8, the frequency of intercourse dips, then rises again when estrogen reaches its monthly high around ovulation. This happy conjunction of the desire for intercourse and fertility is perhaps the result of an inborn program to foster survival of the species.

There is another upturn in sexual activity a few days before menstruation starts; in this instance, the surge in sexual interest is thought to be related to high levels of progesterone and may reflect a biologically driven search for a way to reduce the pelvic congestion that progesterone induces. The rapid involuntary spasms of the pelvic region that occur during orgasm may accomplish this goal.

RHYTHMS DURING MENOPAUSE

Menopause, or the cessation of menstruation, marks the end of a woman's reproductive years. For about three or four years approaching menopause, the ovaries produce eggs less regularly, causing cycles to lengthen. The ovaries usually stop functioning altogether between ages forty-five and fifty-five. When hormone production falls below levels required to build up the lining of the uterus each month, periods stop.

The declining level of estrogen may cause unpleasant symptoms, the most prominent of which are called *hot flashes* or *hot flushes*. Hot flashes follow a circadian pattern, typically worsening at night, which has led some researchers to view them as a disorder of the body's temperature-control system—an exaggeration of the swings that ordinarily occur in sleep.

When hot flashes occur, a woman may develop reddening of the face and patchy red areas on her chest, back, shoulders, and upper arms. As her body temperature readjusts, she may perspire, sometimes so copiously as to drench her clothing and cause chilling. Sometimes the heat and sweating are accompanied by palpitations and a feeling of anxiety. Hot flashes may last from a few seconds to a half hour or longer. The sensations vary from woman to woman and from one episode to the next. Two thirds of all women experience hot flashes during menopause. For most, they prove a modest discomfort, but for one woman in ten, more than three million American women, hot flashes may occur once or more in an hour, disrupting daily activities, as well as sleep, for a year or longer.

In a study of sleep quality in women of menopausal age, Joan Shaver and her colleagues at the University of Washington found that women who experienced hot flashes, not surprisingly, slept worse than those who did not. Age alone did not make sleep worse. Unless they were having hot flashes, women nearing age sixty slept as well as women in their forties.

Sleep disruption may be the underlying cause of other symptoms of menopause, such as irritability and depression. Or depression may cause both poor sleep and other menopausal symptoms, including hot flashes; one study found that women who were already depressed were twice as likely as others to experience hot flashes. Several studies show that estrogen replacement has a domino effect: by reducing the number of hot flashes, it improves both sleep and mood. Other drugs may provide similar benefits. These include progesterone and clonidine, a drug used to treat high blood pressure.

SEXUAL RHYTHMS IN MEN

While there is no monthly cycle in men comparable to the menstrual cycle, male fertility does involve a regular cycle of sperm production. The life of a sperm cell is about two and one half to three months. Sperm are produced in the testes and pass from each testis into a long, coiled tube, the epididymis, where they mature and remain until shortly before ejaculation. At that time, they are propelled into a long duct called the vas deferens and from there to the seminal vesicles, a pair of sacs behind the bladder that produce seminal fluid; this substance, when combined with sperm, becomes semen. Semen travels from the seminal vesicles to the urethra, passing through the prostate gland, where additional fluids are added. During orgasm, semen is ejaculated through the erect penis via the urethra.

Sperm production has a yearly rhythm, with peaks normally in the spring and fall. Light may influence this rhythm, just as it may the menstrual cycle. Animal studies show clearly that day length serves as a synchronizing signal for the mating season, regardless of whether mating occurs in spring or fall. That may be true for humans, too. Day length is a more important factor than temperature in sperm production because it is more consistent. For example, a hot day could occur in December or a cold day in July, but the length of daylight for that time of year will stay the same.

Richard Levine of the Chemical Industry Institute of Toxicology in Research Triangle Park, North Carolina, and his colleagues studied men who worked outdoors in the hot and sunny climate of San Antonio, Texas. They found that the semen quality of the men was lower in the summer than in the winter. The researchers initially suspected that heat might have caused the problem, since heat is known to lower sperm counts, and all of the men worked outside at least four hours daily, but they found no differences between men who spent part of their summer days in an air-conditioned environment and those who did not. Melatonin may account for the summer/winter difference; in the longer days of summer, less melatonin is secreted.

MALE HORMONE RHYTHMS

The chief male hormone, testosterone, is responsible for the development of male sex organs at puberty, and for such male characteristics as the growth of facial and body hair, deepening of the voice, and increase in muscle bulk. In adult men, testosterone plays a key role in sexual drive, performance, and fertility. Testosterone also stimulates aggressive behavior and competitiveness.

Testosterone secretion in men has a daily rhythm, which varies over the year: it is highest in the early morning in May and in the early afternoon in November, a difference of more than six hours, presumably in response to seasonal changes in light intensity. The difference between the daily high and low may be as much as 30 percent. The amount of testosterone produced varies seasonally. The peak is in October, which may explain why October is also the peak month for sexual activity and conception. The low occurs in March, a difference of about 25 percent. Much hoopla has been made about a baby boom on the East Coast of the United States nine months after the largest-ever electrical blackout hit New York, Massachusetts, seven other states, and parts of Canada on November 9, 1965, but the July 1966 birthrate in fact was no different from usual for that time of year; reflecting the annual testosterone rhythm, the late summer and early fall months are almost always peak times for births.

Sexual activity has a daily rhythm, but it appears to be one of convenience, not biology. Alain Reinberg, of the Rothschild Foundation in Paris, and Michel Lagoguey, of the University of Paris, conducted a study in which five healthy young adult Parisian men recorded each time they had intercourse or masturbated for fourteen months, noting the hour and day. The volunteers, who were medical students or biochemists, kept regular hours for going to bed and arising to help keep their daily rhythms synchronized. Their most favored time for intercourse was 11 P.M., with the hours between 6 and 9 A.M. in second place. Not surprisingly, the least likely times were mealtimes and the usual hours of sleep. In this study, the peak months for both sexual intercourse and masturbation were October and, to a lesser extent, July and September. The men had intercourse more than twice as often in October as in February.

THE MYTH OF MALE MENOPAUSE

About 5 percent of men in their forties and fifties have what is commonly called a "midlife crisis," a *Sturm und Drang* experience that may prompt them to change jobs, or leave their spouses and families, or at least wrestle with major questions of self-identity. The midlife crisis is popularly blamed on an alleged "male menopause." But from a biologic perspective, the male menopause is fictive.

While the production of testosterone in men falls with aging, the decline is gradual; it is not analogous to the dramatic plunge in estrogen that occurs in women during menopause. At about age forty, the daily ups and downs in testosterone secretion start to level out. By about age seventy, men may produce 40 percent less testosterone than they did in their forties, but unlike older women, men in their seventies may still be fertile.

Lowered levels of testosterone are thought to contribute to certain physical changes that occur with aging, including an increased tendency to gain weight, lowered bone and muscle mass, and lowered stamina and fertility. There is no evidence that hormone-replacement therapy—testosterone pills or injections—will reverse these changes. Moreover, it may increase a man's risk of prostate cancer.

USING BIOLOGIC RHYTHMS TO IMPROVE MALE SEXUAL PERFORMANCE

Chronobiology research is improving the diagnosis and treatment of men who have trouble getting and keeping erections, a problem that was until recently—but incorrectly—called *impotence*. The widespread use of the sleep lab to investigate causes of erectile disorders followed a 1976 report by Ismet Karacan and his colleagues at the Baylor College of Medicine in Houston, Texas, that healthy males of all ages, from infants to those in their eighties, usually experienced erections throughout REM sleep.

When these researchers monitored erections in the sleep laboratory by attaching a strain gauge to the penis, they found that healthy men in their twenties had an average of four erections a night, lasting a total of 191 minutes, about one third of the time they were asleep. At age seventy, healthy men still experienced three erections a night, for an average of ninety-six minutes, or about one fifth of their time asleep. This rhythm does not appear to depend on testosterone, since prepubertal boys, who lack it, have the same pattern of erections during sleep as adult men.

Most men experience occasional erectile difficulties, from causes as diverse as fatigue, stress, or a bout of influenza, but the problem is usually short-lived. Some men with chronic problems getting or keeping erections prove to have normal erections during sleep, a finding that points to anxiety, anger, or other psychological factors as the most likely cause of their difficulties during waking hours.

By contrast, the lack of normal sleep erections suggests that physical illness is more likely to be the chief cause of the waking problems. For example, diabetes or high blood pressure may interfere with the nerves controlling blood flow to the penis, or to the blood vessels whose engorgement sustains erections. Side effects of medications taken for these and other illnesses may also cause erectile difficulties; certainly, any man who experiences such problems after starting a new medication should discuss the matter with his physician.

The recent discovery that sleep erections follow a circadian rhythm may improve lovemaking for some men with erectile difficulties, according to researchers at Johns Hopkins University. Using a device that measures penile rigidity, Richard Allen and his colleagues observed that erections late in sleep are stronger than those earlier in the night. Erections that start during the last REM episode often persist into waking. Men who have erectile difficulties could use this to their advantage by setting the alarm clock for an hour or two before their usual waking time and attempting intercourse then.

This discovery, like those illuminating other patterns of sexual functioning, enlarges the picture of humans as rhythmic creatures, both awake and asleep. Documentation of the multiplicity of biologic rhythms continues. The next step is to apply these findings to curing all manner of distressing diseases.

PART II

MENDING BROKEN CLOCKS

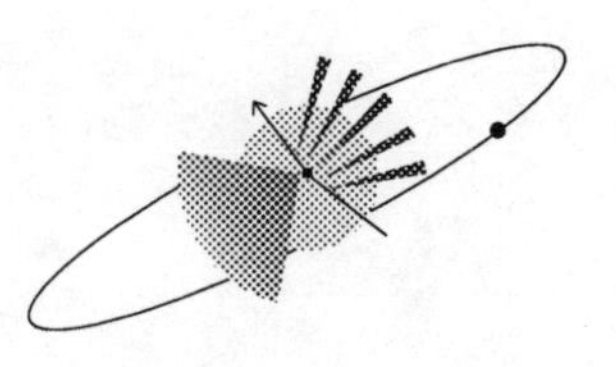

CHAPTER 6

BETTER SLEEP

ONE OF THE MOST OBVIOUS and direct applications of chronobiology research is in the treatment of poor sleep. Difficulty in sleeping is second only to the common cold on the list of Americans' most frequent health complaints. Nearly everyone has an occasional bad night. One out of every three American adults reports suffering a bout of insomnia at least once a year, and about one in ten adults describes the problem as a chronic experience that occurs most nights or every night. Insomnia becomes more frequent as people grow older, affecting some 90 percent of those over age sixty.

Contrary to popular belief, insomnia is actually not a disorder. It is a symptom of several disorders that differ markedly in both origin and treatment. Some are not even problems with sleep per se, but rather with the timing of sleep. People who complain of insomnia may be able to sleep quite well, although the difficulties they suffer when they attempt to sleep typically obscure that fact. Their symptoms arise when they try to sleep at the wrong time on their body clock. Their specific complaint—not being able to sustain sleep, say, or not being able to stay awake in the evening—often provides the critical clue as to which part of the clock has broken down. Chronobiologists have not only identified several distinct disorders of sleep timing, but have also already devised effective treatments for some, including chronobiologically designed schedules, exposure to measured doses of light, and the use of a new drug derived from the hormone melatonin. Following is a report of the latest advances.

THE FIVE MAJOR TYPES OF INSOMNIA

Type 1: Broken Sleep

It would be unusual not to sleep at all, which is what insomnia literally means, although many people lie awake for long stretches. One of the world's "worst things," according to an ancient Egyptian text, is "to be in bed and sleep not." Some people plump the pillow and rearrange the blankets; some stare at the clock, while others turn it to the wall. Counting sheep or bottles of beer remains a favored remedy of many insomniacs; others mentally rearrange closets or subtract from one hundred by threes. An inner voice that orders, "Relax," invariably produces the opposite effect. One cannot bring about sleep on command.

If sufficiently frustrated, people may listen to the radio, or watch television, read, or snack. Some pop sleeping pills in the middle of the night, a bad idea because the sleepiness that the pills induce may well linger on into waking hours. Besides a sense of lethargy and weariness, the daytime consequences of disrupted sleep may include diminished concentration, faulty memory, increased irritability, and, overall, somewhat of a pall on the quality of life. A 1991 Gallup Poll found that chronic insomniacs reported more than twice as many fatigue-related auto accidents as people who did not suffer from insomnia.

Although physical illnesses, such as breathing disorders, may be a cause of insomnia, most tossing and turning stems from stress and anxiety. Some people are natural "worriers." Some seem predisposed to fragile sleep in difficult times, much as others in the same circumstances might suffer indigestion or headaches. A transient problem, such as a fight with a spouse, may cause them a night or two of bad sleep; a longer-lasting problem, an impending divorce, say, may make insomnia persist, too. And once insomnia has become a habit, according to Peter Hauri of the Mayo Clinic, it may last even after the problem that triggered it is resolved. People with this type of insomnia may benefit from learning techniques of stress management, relaxation, and of dealing with their worries at times other than bedtime. (For self-help strategies, see the Appendix.) Sleeping pills may help anxious insomniacs to fall asleep sooner and stay asleep longer, in part because such drugs also reduce the fretting associated with broken sleep. But these medications are designed for short-term use, generally for no more than three weeks, and they rapidly lose their effectiveness.

People who develop insomnia in stressful situations often worry that not getting enough sleep will have some dire effect on their health. In their eagerness to find a remedy, they may unwittingly throw their sleep clock further off balance. Hoping to catch up with lost sleep, they often go to bed extremely early. This is the wrong thing to do. Instead of sleeping longer, they often find they doze and awaken frequently through the night. An insomniac who goes to bed at 10 P.M. and gets up at 7 A.M., may sleep only five or six of those nine hours. Multiple awakenings produce the subjective impression of not having slept at all, leading the person to swear, "I didn't sleep a wink last night."

Why is it that sleepy people do not fall asleep easily when they go to bed early? New understanding of the existence of sleep gates solves this mystery. Studies of the ability to fall asleep across the day by Peretz Lavie and his colleagues at the Technion-Israel Institute of Technology have shown that sleep comes easily at some times, and with difficulty at other times. (See also Chapter 4.) The gates to sleep are closed approximately one to three hours before one's usual bedtime. Insomniacs who go to bed early, already anxious about not sleeping, become even more panicky, thus perpetuating their problem.

Observation of insomniacs' sleep habits has led to a seemingly paradoxical theory: to get more sleep, insomniacs should spend less time in bed. "Reducing time in bed will consolidate sleep," Arthur Spielman of the City College of New York has noted. He and his colleagues at New York's Montefiore Hospital have devised a novel treatment they call *sleep restriction.*

In their studies, insomniacs first limited their time in bed to the total hours they had actually been sleeping before the sleep-restriction program began. Those averaging only five hours of sleep spent only five hours in bed, and six-hour sleepers, six hours. Five hours proved to be the lowest amount people found tolerable, even if they had been sleeping even less, Spielman said. Most people on sleep restriction had to go to bed much later than usual. A five-hour sleeper who wished to get up at 7 A.M., for example, had to go to bed at 2 A.M. The study participants were asked to get up at the same time every morning.

After a week on this regimen, and each week for the next four weeks, the five-hour sleepers added thirty minutes to the time they spent in bed, and the six-hour sleepers added fifteen minutes. By the end of the month, most could sleep seven hours a night, and most reported that their sleep quality was much improved. During the study, the subjects discussed their progress and problems frequently with the researchers; this psychological support, the researchers suggested, boosted the subjects' motivation to stick with the rigors of the therapy. Sleep restriction can work for do-it-yourselfers, but it requires considerable diligence. Despite good intentions, most people find it hard to stay out of bed until the middle of the night; one needs ways to occupy time during the lonely early-morning hours.

While most people with insomnia focus on their difficulty sleeping, insomnia is in fact a twenty-four-hour disorder. One natural repercussion of sleeping poorly is feeling too weary to undertake many physical activities in the daytime. This is a Catch-22 situation, however: people who lead sedentary lives often develop insomnia, and a sedentary lifestyle may perpetuate a sleep problem.

"Since sleep is controlled by the balance within the body of wakefulness and sleep," Richard Allen of Johns Hopkins University has noted, "increasing alertness during the day should have the reciprocal effect of increasing sleepiness at night." In a year-long study based on this chronobiologic principle, fifty-one insomniacs followed a treatment regimen designed to increase their daytime alertness.

Tapping into the activating effects of sunlight, Allen advised the insomniacs to spend at least twenty minutes outside first thing in the morning. He suggested they get at least fifteen to thirty minutes of strenuous daily exercise, such as jogging or squash, in the late afternoon, and an equal amount of moderate exercise, such as walking or stretching, about two hours before bedtime. At the end of the year, the thirty-nine participants who stuck with this regime reported that they slept much better. (For more on exercise and other strategies to improve sleep, see the Appendix.)

Controlled exposure to artificial bright light also may help strengthen twenty-four-hour rhythms in insomniacs. Michael Terman and his colleagues at the New York State Psychiatric Institute developed a computer-controlled lighting device that simulated sunset and sunrise, with light that gradually faded or grew brighter at the appropriate times of day. After just one to two weeks with this treatment, insomniacs who previously needed an hour to fall asleep consistently sank into slumber within thirty minutes; they also started their day feeling more alert, Terman and his colleagues reported. The device would be most useful in cloudy, northern climates in winter, when hours of daylight are limited.

Type 2: Falling Asleep Too Late

Difficulty falling asleep until 3 or 4 A.M. or even later may stem not from anxiety, but from having a sleep clock that is several hours out of sync with that of the rest of the world. In contrast to those whose minds race at bedtime and who awaken frequently, people with this problem sleep quite well. If left undisturbed, they would sleep a normal amount of time, although that might be until noon. Few jobs or schools, however, permit such latitude. What is more likely is that the people who drag themselves out of bed at 7 or 8 A.M., as drowsy as anyone would be after only three or four hours of sleep, are viewed by many as lazy or unmotivated. Before being diagnosed with this disorder, one U.S. Marine faced a court-martial for his "failure to go."

In chronobiologic terms, people with this problem are "out of phase," that is, they are on a different schedule from everyone else. This disorder is called the *delayed sleep phase syndrome*, or DSPS. Although DSPS is infrequently diagnosed in the average physician's office, sleep specialists estimate that it affects 10 percent of all people who have trouble sleeping.

Most people with DSPS prove to be extreme owls on the Owl/Lark Self-test (see page 50). Still, for most, "morningness" or "eveningness" is a preference, not a disability. Most owls can adapt to earlier hours when required to do so. Those suffering from DSPS say that they cannot. According to Richard Ferber of the Children's Hospital in Boston, the problem often first appears during adolescence when youngsters start going to bed late and, more significantly, sleeping late to avoid confrontations with their parents. People living in extremely northern latitudes above the Arctic Circle develop DSPS more frequently than those in more southern climes. The dark period of the seasonal light/dark cycle seems to act as a trigger, presumably by eliminating the opportunity to use light as a zeitgeber.

Some adults with DSPS date their troubles to late-night studying or partying, or to working on the evening or night shift. They say that when they attempted to go back to a more conventional schedule, they were unable to do so. Certain personality factors may also come into play: a need to assert independence or to flout authority, for example. People with DSPS are more likely to blame the world and its "frivolous" rules than to acknowledge that their own behavior contributes to the problem. They may insist that they cannot give up late-night television viewing, for instance, even though they recognize that their sleep schedule interferes with the rest of their lives. One clever student managed to program his university's computer to schedule all of his classes in the afternoon.

People with DSPS are not free-running; if they were, they would move continually around the clock. Numerous zeitgebers, including family, school and work responsibilities, mealtimes, clocks, and the natural alternation of light and dark, keep them moored in a twenty-four-hour world, albeit at the wrong place. Attempts to go to bed several hours earlier prove unsuccessful; the sleep gates stay locked. Nor do anxiety- and stress-reducing techniques or sleeping pills help, because they do not address the disorder of timing that is central to DSPS. These people need a way to reset their sleep clocks.

For a start, sleep experts advise people with DSPS to arise at a conventional wake-up time seven days a week—hard to do, at least in the beginning, because this regimen produces extreme sleepiness during the day. However, that is exactly the point. The hope is to make DSPS sufferers feel sleepy enough to fall asleep somewhat earlier than usual—perhaps an hour earlier each day—and thus gradually to turn their clocks back.

A more drastic treatment, *chronotherapy*, shifts the entire sleep period forward around the clock until more desirable bed- and wake-up times are reached. This treatment puts to advantage the discovery that most people can adapt reasonably well to days as long as twenty-seven hours. While undergoing chronotherapy, which takes about a week, a person would go to sleep and get up about three hours later each day. On the first day, a seven-hour sleeper whose habitual bedtime is 4 A.M. would go to bed at 7 A.M. and arise at 2 P.M. The second day, bedtime would be 10 A.M., and the third day, 1 P.M. On reaching the target bedtime, 1 A.M., say, with a wake-up time of 8 A.M., the person would be expected to stick to it every day.

Chronotherapy requires a protected, quiet environment because many hours of sleep will occur during the daytime. It is also beneficial to have a companion to help the person stay awake until the designated bedtime. One would not want to start chronotherapy and stop in the middle; that would be like traveling halfway around the world and not going any farther. Thus, this treatment is usually conducted in a sleep laboratory.

Sleep experts have been trying to devise shortcuts for chronotherapy. One approach, which packs the entire circuit around the clock into a single weekend, has proven particularly popular with adolescents, who tend to regard its rather rigorous requirements as an adventure akin to an Outward Bound experience. It is also more successful with young adults because their circadian rhythms are more adaptable than those of older people. On Friday night, they must skip sleep entirely; on Saturday night, their big sleep debt should help them to fall asleep at least one to two hours earlier than usual. The next requirement is to arise on Sunday at 7 A.M., and to do so every day from then on, while trying to go to sleep at midnight every night.

Once these people are back on a normal schedule, exposure to sunlight or bright artificial lights in the morning may help them to stay in sync. In one study, conducted by Norman Rosenthal and his colleagues at the National Institute of Mental Health and Johns Hopkins University, twenty people with DSPS were able to maintain their desired schedules when they got up at the same time every morning and sat in front of a light box for two hours. The study participants also wore dark goggles between 4 P.M. and dusk to diminish the alerting effects of light late in the day.

Chronotherapy is a last resort for those whose clocks are way out of alignment. It is not without hazards. When cut loose from a twenty-four-hour day, some people find that they cannot go back to it. It is as if the hands on their clocks start spinning wildly. Although they had previously accommodated to living out of phase, they may be impelled by chronotherapy into non–twenty-four-hour rhythms. (See section on this problem, page 104.)

Type 3: Falling Asleep Too Early

The clock can also get stuck in a "too early" position. People with the *advanced sleep phase syndrome*, or ASPS, become overwhelmingly sleepy much sooner than usual. They may choose to go to bed at 7 or 8 P.M., sometimes even earlier. Understandably, even if they sleep a normal amount of time, they still awaken in the middle of the night. Then they become distressed about being awake when they "are supposed to be asleep."

ASPS generally occurs in people over sixty-five. With age, the circadian clock speeds up, and most people become more larkish. The low point of body temperature in people sixty-five and older occurs about two hours earlier than in younger adults. Since people tend to awaken as temperature rises, it is no wonder that older people awaken earlier and that, as a result, they go to sleep earlier. Most adapt satisfactorily to this change, although they may find it an inconvenience. In one survey, one third of a group of older patients told their doctors that they awakened too early nearly every day. In those with ASPS, the early-to-bed, early-to-rise lifestyle is extreme. More is at stake than the embarrassment of falling asleep at dinner parties; falling asleep at the wheel on the way home may be fatal.

When awakening in the early-morning hours, a person may find it hard to pursue ordinary waking activities while others sleep. At 3 A.M., a household is easily disrupted by the smell of coffee or the sound of someone tiptoeing around. Consequently, some people linger in bed or take sleeping pills in the hope of extending their slumber. Unfortunately, the pills may increase sleepiness during the day, promoting napping or an even earlier bedtime the next night. Even if they force themselves to stay awake until a conventional bedtime, their internal wake-up alarm may still sound at 3 A.M., which in turn may cause them to become even sleepier during the day than on their former schedule.

As with DSPS sufferers, light/dark cues and social cues usually keep people with ASPS locked into somewhat of a normal day. They do not keep going to bed earlier and earlier. Chronotherapy that goes backward around the clock would seem a logical treatment. Because it fights the body's natural proclivity to extend the day, however, it is harder to accomplish in three-hour increments than is the case for DSPS. Because an hour or so a day is all that most people can manage, chronotherapy for ASPS may take a month. Hence, it is not practical.

For many, the most effective strategy is to take a thirty-minute nap in the late evening, get up, and stave off bedtime until approximately midnight. This tactic delays the middle-of-the-night awakenings, shifting the time of arising to a more acceptable hour.

Proper timing of light exposure may also be beneficial. Exposure to bright light in the evening may help counter early-evening sleepiness and solidify nighttime sleep, studies by Scott Campbell of the New York Hospital-Cornell Medical Center and his colleagues have shown. One can avoid the alerting effects of morning light by sleeping in a dark room, keeping the shades drawn until noon, and wearing dark sunglasses if going outdoors. Charles Czeisler of Harvard and his colleagues helped a sixty-six-year-old woman who, though living in Boston, had a body clock that ran on London time. They used light therapy to bring her back home.

Type 4: Disorganized Sleep/Wake Schedules

Most people have experienced being in bed with the flu or a bad cold, sleeping on and off without paying much attention to the time of day. This pattern, referred to as an *irregular sleep/wake pattern*, sometimes becomes chronic, particularly in elderly people who take daytime naps to compensate for their reduced ability to sustain sleep at night. This sometimes creates a vicious circle: the more they nap during the day, the worse they sleep at night. And the worse they sleep, the more they nap.

A decline in daily zeitgebers, particularly social contacts, has been widely blamed for the development of irregular sleep/wake patterns in older people. But a recent study suggests that this belief is wrong. Timothy Monk and his colleagues at the Western Psychiatric Institute and Clinic in Pittsburgh found that their subjects, forty-five persons over age seventy, showed greater regularity in their daily social rhythms than did a comparison group of people in their twenties. The older adults were more likely to have companions with whom they shared meals, exercised, and watched television, and their days were equally full with a variety of activities. Indeed, the seniors were far more likely than the younger group to stick to a fixed routine. Nonetheless, their sleep was considerably worse. What accounted for this discrepancy? It is not only sleep that deteriorates with age, the researchers said. The ability to recognize and act on zeitgebers may also diminish, somewhat like the ability to hear a whisper.

In related research, also at the University of Pittsburgh, behavioral neuroscientist Melanie Kittrell serves as manager of a posh rat "hotel," where she is seeking to discover whether old rats with young roommates have more stable daily rhythms than old rats with old roommates, or with no roommates. The animal research may illuminate further the impact of social influences on rhythms in the elderly.

In healthy older people, regular routines—particularly rising at the same time each morning, restricting sleep to six and one half to seven and one half hours, and keeping naps to a minimum—appear to help strengthen nighttime sleep. Unfortunately, when illness comes on the scene, regular habits may fly out the door. One seventy-one-year-old man had to stay in bed after breaking his leg. He had few visitors and spent most of his hours watching television. In a short time, he began to sleep intermittently through the day, and even after his bones healed, the habit persisted.

In extreme cases, usually associated with chronic illness, people may drift into a pattern of eating and sleeping that is almost like that of newborn babies, except that babies usually get more sleep. In a study of more than two hundred nursing home residents, Sonia Ancoli-Israel of the University of California at San Diego found that the majority were never asleep for a full hour, nor were they ever awake for a full hour, throughout the twenty-four-hour day. This startling and disturbing finding implies an enormous disability, severe enough to interfere with visits from family and friends and enjoyment of simple activities such as watching television. A huge number of people may be affected; two thirds of Americans over age sixty-five will spend part of their lives in long-term care facilities.

The problem may be compounded by admission to a hospital, particularly an intensive-care unit where lights are on constantly. Happily, this is one area of the hospital where chronobiology has made some headway. To keep patients from losing track of whether it is night or day, patient rooms increasingly have a large round standard clock installed on the wall at the foot of the bed, the kind of clock familiar to most adults from their childhood schoolrooms. Nurses are instructed to walk into a patient's room when they first come on duty and say, "Good morning," or "Good afternoon." In rooms with windows, curtains are opened and closed at the right times as reminders of the normal cycling of day and night.

Further, federal law requires that all hospital rooms constructed or remodeled after 1977 include a window or skylight. When a wing added after that date to the Stanford University Medical Center blocked the view from three rooms in the cardiac intensive-care unit, the hospital found a way to compensate. It installed artificial windows containing larger-than-life full-color photographs of lush, green countryside. Behind the photos are computer-controlled light boxes that change 650 times a day, mimicking natural light from sunrise to sunset.

People who suffer from Alzheimer's disease, Parkinson's disease, or the effects of a stroke seem particularly prone to experience *sundowning*, a term used to describe confusion, disorientation, and agitated behavior that occurs after nightfall. As many as 20 percent of demented persons are estimated to suffer from sundowning. Some wander around their homes, shouting and breaking things, and some even go out of doors barefooted and wearing only nightclothes. This disruption in daily rhythms puts a severe strain on families and caregivers, often disturbing their own sleep, and it is the chief factor cited in their decision to institutionalize an older person, according to a study by Charles Pollak and Deborah Perlick of the New York Hospital-Cornell Medical Center. The problem then is merely transferred to the nursing home staff, who are apt to deal with it by increasing sedation. People over age sixty-five comprise 11 percent of the United States population, but they receive 40 percent of the prescriptions for sleeping pills.

Sundowning has been observed to be more severe in winter. The lower level of natural light at that time of year may make it harder to maintain robust twenty-four-hour rhythms. The use of bright lights is being investigated as a possible way to shore up failing rhythms; in one study, Andrew Satlin of Harvard University and his colleagues found that exposure to two hours of bright light in the morning reduced the nocturnal restlessness of patients with Alzheimer's disease. At present, caregivers can best minimize sundowning by keeping the person awake and occupied during the day. This requires a regular activity program including morning and afternoon exposure to outdoor light and exercise, such as walking, if the person is up to it. Caregivers may find that the patient's sleep improves if they eliminate caffeine from the person's diet, and establish a bedtime ritual, such as listening to music, brushing hair, or having a light snack. A night-light may help reduce confusion if the person awakens during the night. The potential benefit from sleeping pills has to be balanced against the likelihood of increasing daytime sleepiness because the rate of metabolism and excretion of these drugs slows with age.

Type 5: Non–Twenty-Four-Hour Rhythms

Unlike the great majority of the population, some people are unable to reset their inner clocks to a twenty-four-hour schedule day after day. The beat of their internal drummer insists on being heard. In contrast to those with irregular sleep/wake patterns, people with this problem, known as the *non–twenty-four-hour sleep/wake syndrome,* have relentlessly orderly clocks. Indeed, their clocks act as if they were free-running in a chronobiology laboratory. Even though these individuals live in a twenty-four-hour world with abundant time cues, and despite their attempts to adhere to a conventional schedule, their sleep phase travels in and out of synchrony with the usual hours for sleep. Thus, they periodically suffer insomnia at night, and sleepiness during the daytime.

This disorder was first described in 1977 in a man who had been blind since birth. Jim Stevenson, then twenty-eight, a biostatistician and experimental psychologist with a Ph.D. from Stanford University, said that he had struggled for his entire life to keep regular hours. To do so, he often required sleeping pills at night and stimulants during the day.

Initially, Stevenson's problem was thought to be extremely rare. That has turned out not to be so. Study after study has revealed that perhaps three out of every four blind persons suffer from this disorder. It also occurs occasionally in people with normal vision. Until chronobiologists came on the scene, the fact that sleep complaints were common in blind people, and that such complaints were cyclic in nature, went unnoticed.

The difficulties that blind people experience provide a dramatic illustration of both the importance of light and dark in setting human body clocks, and the surprising weakness of other cues—including regular hours for sleep and meals—in their absence. Studies of circadian rhythms in blind people promise to advance our understanding of how inner clocks work, and thus will benefit both the sighted and the blind. Further, the recognition of the link between blindness and rhythm disorders has helped make physicians who treat eye injuries and diseases more aware of the importance of preserving as much residual visual input as possible. People who can distinguish even low levels of light can entrain themselves to a twenty-four-hour world far better than those with no light sensitivity at all.

Torsten Klein and his colleagues at Harvard Medical School studied sleep patterns and rhythms of body temperature, urinary excretion, and release of the hormone cortisol in a blind man with chronic non–twenty-four-hour rhythms. This man, who stayed in their chronobiology laboratory for 124 days, lived on a regular twenty-four-hour schedule, going to bed and getting up at the same time each day. Nonetheless, his body rhythms continued to have periods that exceeded twenty-four hours by twelve to sixteen minutes. This seemingly small difference had big repercussions: when his sleep was disturbed, for instance, he would awaken twice a night to use the bathroom. This type of urinary frequency is consistent with normal wakefulness. When his temperature was elevated, on a par with normal waking levels, he slept restlessly.

Laughton Miles and his colleagues at Stanford University, the scientists who initially documented Stevenson's problem, had imposed a rigid schedule in hopes of anchoring his clock more firmly. Stevenson went to bed at 11 P.M., and arose at 7 A.M. He ate meals at the same time each day, and avoided naps. Nonetheless, he became progressively more restless at night, and more sleepy during the day.

Over the years, Stevenson had tried a vast array of measures to improve his life. These included cold showers, loud music, and exercise in the morning, and foods alleged to promote alertness or sleepiness at the appropriate times of day. He had even surrounded his bed with a halolike device to neutralize the earth's magnetic field, a tactic that had successfully regularized circadian rhythms in sparrows. All of his efforts proved futile.

More recently, Stevenson acquired a constancy he had never before known, thanks to the hormone melatonin, which appears to play a role in inducing sleep. In sighted people, melatonin secretion ordinarily starts about 9 P.M. and continues until daylight. In totally blind people, melatonin free-runs—that is, the period of secretion moves around the clock. Psychiatrist Robert Sack and his colleagues at the Oregon Health Sciences University found that blind people who slept in sync with their melatonin rhythm slept just fine. This prompted them to offer melatonin to a group of blind people, who took it at their desired bedtime when the bedtime and melatonin rhythms were in sync. The happy result was that their melatonin cycle locked in to this schedule, relieving both their insomnia and daytime sleepiness.

Stevenson was a participant in Sack's studies. In the first year that he used melatonin, Stevenson said, he needed to take a sleeping pill only 6 times, whereas he had done so about 250 times the year before. He slept well for about eight hours and, most significantly, at the same time every day. "My social life is much better, since I now can count on being able to do things in the evening," he related. "With melatonin," he summarized, "my whole life has changed." Small doses of melatonin, a substance that the body produces naturally, hold the tantalizing promise of being able to reset rhythms—a big improvement over any drug presently available. However, as of this writing, melatonin was still regarded as an investigational new drug, and hence is restricted to research programs.

Curiously, physicians have known for centuries that susceptibility to many illnesses, and the signs and symptoms of many illnesses, come and go at different times of the day and the year. Aristotle noted that seizures often begin in sleep. Hippocrates said, "Whoever wishes to pursue the science of medicine in a direct manner must first investigate the seasons of the year and what occurs in them." Over the years, numerous physicians have suggested that disruptions in bodily harmony may predict illnesses, and perhaps may even cause them. But because many illnesses—heart attacks, strokes, and episodes of depression, for example—occur only once or just a few times in any person's life, most physicians gave little thought to the significance of body time. The benefits that have been demonstrated in treating sleep disorders according to chronobiologic principles imply that a big revolution in medical care is nearly at hand.

CHAPTER 7

THE RHYTHMS OF SICKNESS AND HEALTH

CHRONOBIOLOGY IS SUCH A YOUNG SCIENCE that most physicians practicing today did not study it in medical school, and many still do not know much about it. There is always a time lag, of course, between discoveries made in the laboratory and their application to patient care. The first scientific paper in chronobiology, the report of de Mairan's 1729 study of plant behavior, somberly concluded: "The progress of true Physics (Science), which is experimental, can only be very slow." An even more significant reason for the slowness in acceptance of chronobiologic principles is that they pose a major challenge to the medical establishment. Compared with chronobiology, the majority of advances in medicine, even so-called "breakthroughs," are merely baby steps forward. Chronobiology is a radically new way of conceptualizing health and disease.

To adopt chronobiology, physicians would have to discard much of the long-enshrined principle of *homeostasis*, or self-regulation. The word "homeostasis," from the Latin words *homeo*, meaning the same, and *stasis*, state, was created in 1925 by the renowned Harvard Medical School physiologist Walter B. Cannon. The theory behind the principle had originated in 1885 with the French physiologist Claude Bernard, who suggested that the body strives to keep its *milieu interieur*, its inner environment, as constant as possible to shield it from the assaults of the outer world.

Extending Bernard's work, Cannon suggested that ups and downs in various bodily functions represented the body's fine-tuning within a fixed range to maintain an equilibrium, which he called a "steady state." Homeostasis explains why, for instance, people sweat when they are hot and shiver when they are cold: an internal temperature control system keeps the body from overheating or overcooling.

Cannon observed variations in blood pressure, temperature, and even in concentrations of hormones in the blood. But he did not see the time-of-day patterns in these variations nor did he recognize their importance. Cannon believed the changes he saw were random, and significant only when people were sick. Widespread acceptance of homeostasis explains why a body temperature of 98.6°F is commonly referred to as "normal," when it is merely an average. As the chart, "The Daily Temperature Cycle," on page 35 shows, a reading of 98.6°F at 4 A.M. indicates a fever.

Chronobiology studies have shown that blood pressure, hormone levels, urine volume, and many other measures of bodily performance not only fluctuate within a substantial range over the course of a day, week, month, or year, but also often deviate well beyond the range associated with disease. Although Bernard and Cannon were right—checks and balances are necessary for survival—there is no true constancy in the body. Some substances present in the blood or urine at one time of day are virtually absent at another. Nevertheless, many, perhaps most, physicians view such fluctuations as simply the effects of sleep, activity, or diet. They regard these changes as "noise" in the system, a view that has served as a powerful braking force to the development of chronobiology.

As early as the 1950s, researchers mounted a challenge to the steady-state hypothesis by demonstrating that rhythms literally could make the difference between life and death. Franz Halberg and his colleagues at the University of Minnesota showed that a fixed dose of a potential poison would kill nearly all mice that received it at one time of day, but would kill only a few or even none of those that received it twelve hours earlier or later. The researchers also showed that loud noise caused mice to have seizures at one time of day but not at others, and that the damage caused by a given dose of X rays depended on when the animals received it.

These findings can be applied directly to humans. Drugs that can help at one time of day may harm at another. Indeed, *when* one takes a medicine may be as critical as *which* medicine one takes. There is often considerable variation over the day in the way the body absorbs, uses, and disposes of drugs, and the way drugs interact with other drugs. It is impossible to achieve a steady state of most drugs in the bloodstream.

Yet physicians typically prescribe medications to be taken in equal doses throughout the day, a practice that dismays and infuriates chronobiologists. Decrying "the stupidity of three times a day drug administration," Alain Reinberg of the Rothschild Foundation in Paris has called this custom "obsolete and in some cases inappropriate or even dangerous."

Chronobiologists assert that drug doses should be adjusted to meet the differing needs or functions of target organs or tissues at various times of day.

Some examples: cancer medication given at the time of day when it will do the most harm to cancer cells and the least damage to normal cells may improve survival. (For details, see pages 117–120.) Allergic rhinitis—hay fever—is a morning disorder. People who awaken with stuffiness or a runny nose could take medicine to relieve their symptoms, but it would be even better if they could prevent the morning symptoms by taking medication before going to bed. The treatment of short stature with growth hormone is most successful if it is given at night, in sync with the hormone's normal time of secretion.

Rhythms that are longer than one day also need to be factored into treatment plans. Take, for example, weekly rhythms. In preantibiotic times, doctors noted that crises in pneumonia occurred on the seventh day, and deaths in malaria came most frequently on days 7, 14, and 21. A study of nearly 150 kidney transplant patients showed that the new organ was rejected more often on days 7, 14, and 21; drugs, such as cyclosporin, that suppress the immune system and boost the chance of a transplant's acceptance may work better when given in higher doses on the days of high rejection risk than in equal doses every day.

The timing of radiation therapy and surgery may alter the benefits gained from these procedures. In a study of patients receiving radiation therapy for mouth tumors, Franz Halberg and his colleagues in India found that people who were treated when their tumors were at a daily temperature peak improved the fastest and had the fewest recurrences. When Halberg needed bypass surgery himself, he calculated the best time not only on his own biologic clock but also on that of his surgeon, taking into account his own blood pressure pattern and his surgeon's sleep schedule. The operation took place early in the morning; both patient and doctor came through just fine.

Medical diagnoses also need to take rhythms into account. For example, white blood cell counts are lowest in the morning and highest in the evening. They vary by 30 percent over the day. Many of the body's two hundred hormones are secreted episodically. Cortisol, for instance, is abundant in the morning (peak secretion is around 7 A.M.) but near zero around midnight, a three- or fourfold difference. The detection of diseases involving over- or underproduction of cortisol requires knowledge of the time a blood sample was obtained. In many illnesses, single samples of blood or urine taken at arbitrary times of day may yield a much higher proportion of false positives and false negatives.

The weight of evidence from animal and human studies would seem hard to brush aside. And, indeed, as experience with a number of diseases described in this chapter suggests, the new ideas are beginning to catch on. But they have not yet become part of mainstream wisdom. Overturning homeostatic precepts remains an uphill battle. "I have twenty-five things to think about in planning treatment," a prominent researcher at one of the nation's leading cancer centers recently declared. "Circadian rhythms are not even on the list."

They should be at the top of the list, chronobiologists say. "Homeostasis is like looking at a house from the outside and trying to figure out what is going on in the house by watching what goes in the door or out the chimney," Halberg has asserted. "Chronobiology goes inside the house."

A DAY IN THE LIFE

Some four hundred years ago, the English writer Robert Burton presciently observed, "Our body is like a clock; if one wheel be amiss, all the rest are disordered, the whole fabric suffers: with such admirable art and harmony is a man composed." Events from birth to death cluster at specific times on the body clock. Following are some of them:

Midnight to 4 A.M.: This is the prime time for racing the stork to the hospital, the hours in which spontaneous labor is most likely to begin, according to a study of nearly 300,000 pregnant women.

Between midnight and noon, natural births occur three times more often than between noon and midnight. Early in human evolution, being born at night or in the early morning may have been advantageous, since such births probably would have occurred in a protected place. Today, this rhythm may actually work against survival. Researchers in Switzerland found that babies born in the evening had a higher death rate than those born during the day, perhaps, they suggested, because fewer staff people were around at night—a daily rhythm but obviously not a biologic one.

1 A.M.: Between 1 and 4 A.M., skin cell division peaks. The notion of needing one's "beauty sleep" is not far-fetched. Cosmetic manufacturers who advertise that skin tissue repair increases in people who wear night cream are telling the truth. Not the whole truth, however, for cell turnover also increases in those not using the cream.

2:30 to 8:30 A.M.—These are the hours during which sudden infant death syndrome (SIDS) strikes most frequently. This disorder, which claims the lives of 5,500 babies in the United States each year, may reflect a breathing abnormality that is undetectable during wakefulness, but is aggravated by sleep and by respiratory illnesses. This theory is supported by the annual rhythm in SIDS deaths. According to the National Center for Health Statistics, twice as many babies die from SIDS in January, a peak month for respiratory illnesses, as in July, the low month for SIDS deaths.

3 to 8 A.M.: These are the hours when toothaches most often start. Sensitivity to all types of pain is highest at this time.

4 to 6 A.M.: These are the most frequent hours of death from *all* causes in people of all ages. The odds of dying at night are about 30 percent greater than dying during the day.

Deaths from all causes are also more common on weekends, on the days immediately before and after one's birthday and on the days after holidays. Deaths from heart attacks, and suicides, are more common on Mondays. The explanation in all cases may be the same: the disruptive impact on biologic rhythms of even minor changes in usual daily routines—insignificant for most people but dangerous for those already ill. Social influences are at work here, too. People often refrain from "bothering" the doctor on a weekend, or disrupting a festive event.

7 A.M.: Aspirin taken at 7 A.M. remains in the body for twenty-two hours. Taken at 7 P.M., it lasts only seventeen hours. An antihistamine taken at 7 A.M. lasts fifteen to seventeen hours, twice as long as it does if taken at 7 P.M. One time of day is not necessarily better than another to take these drugs, but the differences may be important in some circumstances.

8 A.M.: Between 8 A.M. and noon, hay fever symptoms are worse. These are not the same hours that pollens and many other airborne allergy-inducers are highest; that time comes later in the day. If the two periods coincided, hay fever sufferers undoubtedly would be even more miserable.

3 P.M.: People allergic to house dust who inhale it at this hour may suffer only minor symptoms. If they inhale the dust at 11 P.M., they are likely to have severe trouble breathing.

At 3 P.M., lidocaine, a local anesthetic used by dentists, blunts sensitivity to pain nearly three times as long as it does at 7 A.M.—thirty-two minutes in the afternoon compared with only twelve minutes in the morning. The midafternoon may thus be a more desirable time to visit the dentist or undergo surgery. Another plus: motor skills are highest in the afternoon. The dentist's or surgeon's hand will be at its steadiest then.

5 P.M.: Runners have their best performance on the fifty-yard dash at this time. Indeed, peak performance in most sports is achieved in late afternoon and early evening when temperature is highest.

Between 5 and 7 P.M., the senses of hearing, taste, and smell are quite acute. They actually are most acute at 3 A.M. and lowest at 6 A.M, but few people notice.

7 P.M.: Body weight is at its daily high. The low occurs at about 7 A.M.

Between 7 and 11 P.M., itching is most severe in people with the skin disorder atopic dermatitis. These are also the hours of peak daily production of histamine, a chemical that triggers itching and other allergic reactions.

8 P.M.: A martini cocktail drunk in the evening takes longer to get into the bloodstream and has a less intoxicating effect than a Bloody Mary in the morning. In the evening, however, the same amount of alcohol stays in the body longer. (For more on social drinking, see Chapter 3.)

A TIME TO HEAL

Chronobiologic findings have advanced the understanding of some common illnesses and are beginning to have an impact on their treatment.

Heart Attacks and Strokes

Heart attacks and strokes occur more often around 9 A.M. than at any other time of day. This pattern first showed up after the data from three thousand people admitted to the hospital after heart attacks was analyzed. Researchers wondered if these attacks had truly occurred in the morning, or were merely discovered then. Perhaps because they were asleep people did not notice the minor symptoms of pain that signaled an impending attack. In 1985, cardiologist James Muller and his colleagues at Harvard Medical School reported that they had found a way to pinpoint a heart attack's starting time by tracking an enzyme, creatine kinase, that first appeared in the bloodstream about four hours after the episode began. In a study of seven hundred patients, the researchers found that heart attacks started most often between 9 and 10 A.M., and least often, between 11 P.M. and midnight, a threefold difference.

Numerous other studies have substantiated the morning pattern. Although many heart attacks and strokes are not fatal, Merrill Mitler of the Scripps Clinic and Research Foundation and his colleagues found in their review of five thousand death records that people over age sixty-five were more likely to die from heart disease, high blood pressure, or stroke at 8 A.M. than at any other time.

While there is no enzyme test that shows when strokes begin, a study supported by the National Institute of Neurological and Communicative Disorders and Stroke, in which twelve hundred stroke patients, or their families, were interviewed by researchers at four university medical centers, found that in people who were awake when stricken, more strokes occurred between 10 A.M. and noon than in any other two-hour interval. They also discovered that the incidence of strokes declined steadily through the rest of the day, and that strokes were least likely to occur between 10 P.M. and midnight. Some 744 patients in this study suffered a stroke after waking, while 331 awakened with stroke symptoms.

Because the records were not adjusted for the individual's habitual wake-up time or for variable work schedules, the true magnitude of the morning increase may be even greater, the researchers noted. The key factor here, as for all biologic rhythms, is an individual's activity schedule, not clock time. The most dangerous hours for heart attacks and strokes are the first ones after waking, no matter what time of day that is.

Certain other cardiovascular events also occur more often in the morning. People with angina, for example, are more likely to experience chest pain in the morning. Transient ischemic attacks, episodes in which blood flow to the heart slows or stops briefly, occur most frequently within an hour or two after awakening. Although these episodes may not produce any symptoms, they leave their characteristic signature on records of heart muscle activity registered by portable monitors worn around the clock. Such attacks prove to be both more frequent and longer-lasting in the morning, and they are thought to be instrumental in triggering more severe heart attacks later on.

A heart attacks occur when an obstruction in one of the coronary arteries keeps the heart from receiving an adequate oxygen supply. Strokes, which can be thought of as "brain attacks," result from an interruption in the flow of blood to part of the brain. In the majority of strokes, a blood clot blocks an artery, usually one narrowed by fatty plaques; in the remainder of the attacks, the artery bursts, leaking blood onto the surrounding tissue.

Heart attacks and strokes account for about one third of all deaths in the United States each year. About 1 million Americans have heart attacks annually, and 500,000 suffer strokes. Determining why these problems occur most frequently in the morning may help reveal why they happen at all. The last hours of slumber, which are dense with REM sleep, when the heart beats more irregularly and when the breathing rate varies more than at other times of day, may be a more vulnerable time. These stresses present an added challenge to people with angina, irregular heart rhythms, and other heart diseases. Changing from a horizontal to an upright posture, for instance, imposes an enormous demand on the cardiovascular system. The heart pumps blood faster, blood pressure and temperature rise, and blood components called *platelets* increasingly clump together, possibly causing clots to form. Together, these burdens may overwhelm fragile blood vessels in the heart or brain that have been weakened by age or narrowed by disease.

Staying in bed is not the answer. Even closely monitored patients in intensive-care units suffer more heart attacks and strokes in the morning. Any hospital stay, of course, is far from tranquil, and patients get little chance to rest. With doctors and nurses coming and going, X rays being taken, blood being drawn, procedures performed around the clock, and meals provided at unaccustomed times, sleep and other circadian rhythms may undergo vast disruption. Hospitals would do well to heed the morning risk pattern in scheduling work hours, since at present, in most hospitals, the lowest number of staff members are on duty in the early-morning hours.

The morning pattern holds many other implications for treatment. Researchers have found that fasting increases platelet stickiness, suggesting a simple strategy to reduce one's risk of a morning heart attack: eat breakfast. In the future, doctors may advise some patients to take their medications at bedtime, and others, when they awaken in the morning, perhaps even before they get out of bed. Chronobiologists forecast that eventually doctors may prescribe that fast-acting heart drugs, such as nitroglycerin and calcium-channel blockers, be taken in the morning, and that long-acting drugs, such as beta-blockers, be taken at bedtime.

Several major studies have shown that simply taking an aspirin (325 mg) every day or every other day (for men) or one to six times a week (for women) reduces the clumping of platelets, thereby offering protection against heart attacks. The people in these studies did not take aspirin at any specific time of day. Nor in fact would there have been any advantage in doing so: aspirin eliminates morning increases in platelet activity for at least thirty-six hours after it is taken.

Is it a bad idea for people with heart disease, indeed for anyone, to start the day with exercise? Theoretically, morning exercise is more risky because it imposes an additional strain on the heart. To assess the risk, Paul Murray and his colleagues at Wake Forest University compared two groups of people participating in a cardiac rehabilitation program, all recovering from heart attacks, heart surgery, or other cardiac conditions. Both groups exercised three days a week, one group from 7:30 to 8:30 A.M., and the other for an hour between 3 and 5 P.M. All told, the researchers examined the effects of nearly 170,000 hours of exercise on the 221 participants. Only seven cardiac events occurred in the entire group, and time of day made no difference. For people without heart problems who exercise regularly, the risk of suffering a heart attack while exercising presumably is even smaller. The bottom line: according to this study, both times of day are safe and neither time is safer than the other. To eliminate all uncertainty would require a huge number of subjects and many years of observation. The consensus is that the benefits of regular exercise at any time of day outweigh any possible disadvantages. Indeed, after reviewing all available medical literature in 1987, the Centers for Disease Control and Prevention of the United States Public Health Service concluded that a sedentary lifestyle virtually doubled the risk of having a heart attack. Even for weekend athletes, time of day is probably far less significant than the impact of unusual stress.

One can reduce one's risk of heart attack or stroke by not smoking, by controlling blood pressure, maintaining a healthy weight, and consuming a low-fat diet, according to neurologist John Marler of the National Institute of Neurological and Communicative Disorders and Stroke. Most people also benefit from limiting their salt intake. "These are not circadian issues," Marler said, "but they are the strongest preventive advice we now have."

High Blood Pressure (Hypertension)

People commonly think of blood pressure as relatively static, with two numbers, such as 120/80, being a so-called normal reading. But blood pressure rises when one is moving and falls when one is at rest, goes up when one stands and drops when one lies down. Blood pressure is higher when one talks than when one is silent. And, of course, it changes markedly over the day.

Blood pressure starts to rise before one normally awakens, and continues to rise after waking, peaking in late afternoon or early evening. It reaches its low point after sleep starts. A person whose blood pressure is within the purported normal range in the morning may become a candidate for treatment when afternoon readings enter the equation.

The term "blood pressure" refers to the force that the blood exerts against artery walls, as the heart pumps blood through the body. With each beat, or systole, pressure rises, causing the walls of the arteries to stretch. With each rest between beats, or diastole, pressure falls, letting the artery walls relax. In blood pressure readings, the systolic number is written first, as in 130/85.

High blood pressure is the most common chronic cardiovascular disease in the United States; as many as sixty million Americans suffer from it. High blood pressure, or hypertension, has been called "the silent killer" because it may substantially weaken blood vessels before actual symptoms appear. A sustained elevated blood pressure—140/90 or higher—may trigger a heart attack, stroke, or kidney failure.

Even physicians often speak of "the" blood pressure or "the average" blood pressure in a given patient. Chronobiologists say that is a mistake. "Trying to assess high blood pressure with a single measurement or even a series taken at arbitrary times is like taking snapshots of a roller coaster," chronobiologist Halberg has asserted. "Unless we monitor for a forty-eight-hour period and then interpret the data properly, we may treat people who simply have 'white coat' hypertension, a reflection of anxiety from the medical exam," he said. "We may unnecessarily give a patient drugs that render him impotent or give him gout. We may give him a stigma of bad health for life or place his job in danger."

One man routinely saw his doctor on his way to work in the morning. The doctor told him he was in great shape. But when the man's blood pressure was monitored around the clock as part of a research study, it proved to be in the high range 65 percent of the time.

Dramatic changes in blood pressure may emerge only in around-the-clock studies. When monitored while asleep, one man showed a marked rise in blood pressure for nearly an hour, possibly the result of a panic episode in a dream. On another occasion, his blood pressure fell sharply for about an hour. Both episodes were outside the range of normal variability.

New, portable automatic recording devices make it possible to take virtually continuous readings. While such high-tech equipment is typically reserved for one- or two-day diagnostic studies, some doctors are beginning to ask patients under treatment to measure their blood pressure periodically at home to gauge their progress. Home monitoring is easy to do, with inexpensive devices now widely available. Chronobiologists advise taking measurements every hour and even setting the alarm to get a reading in the middle of sleep. One should sit quietly for five minutes before taking a reading. A single high reading should not be viewed as cause for alarm; a regular pattern must be determined.

Many doctors advise that antihypertensive drugs be taken at bedtime to minimize common side effects of the medication, such as dizziness. However, chronobiologists contend that, since blood pressure is ordinarily lower at night, physicians should instead study a patient's profile and prescribe that medications be taken only when blood pressure is characteristically high, not when it already is low.

Chronobiologic findings may help prevent high blood pressure. Halberg and colleagues in Italy and Spain monitored the blood pressure and heart rate of 144 newborn babies every thirty minutes for forty-eight hours. Babies whose families had a history of high blood pressure or related cardiovascular disease showed more daily variability than babies whose families had no such problems. The mother's family proved to be a stronger influence than the father's. This suggests that intervention strategies might begin during pregnancy, with, as one possibility, a low-salt diet for a woman with a family history of high blood pressure.

Moreover, vulnerable children could be taught at an early age to pay attention to their own diet, to exercise regularly, avoid smoking, and adopt other preventive measures to minimize risk of high blood pressure.

Cancer

Drugs that kill cancer cells also destroy normal cells. Ideally, such medication should kill only the malignant cells. New treatment schedules that acknowledge rhythms of both cancer cells and normal cells may bring that goal closer.

So far, researchers have plotted optimal delivery schedules for about twenty of the approximately fifty common anticancer drugs, mainly in animals. Human studies have begun only recently. Certainly, the need for improved treatment is urgent: cancer claims 500,000 lives in the United States annually.

While all cells are more vulnerable to damage when they are dividing, cancer cells typically divide more rapidly than normal cells. Plotting daily rhythms of normal cells may enable doctors to administer anticancer drugs when fewest normal cells are dividing. Timing treatment this way may make it possible both to use higher, potentially more effective doses than would otherwise be safe and to reduce side effects. For example, giving drugs at a time when intestinal cells divide slowly may eliminate the diarrhea that causes many patients to discontinue therapy.

Moreover, some types of cancer cells themselves divide according to a daily rhythm. This may make it easier to target them for treatment. For example, Robert Klevecz and his colleagues at the Beckman Research Institute of the City of Hope in Duarte, California, analyzed cells taken after surgery from the abdominal cavity of ovarian cancer patients in two- to four-hour intervals around the clock. The researchers found that the cancer cells reproduced at twelve-hour frequencies and sometimes even faster. The peak most commonly occurred around 10 A.M. and 10 P.M. These rhythms, Klevecz suggested, "should be exploited for therapeutic benefit."

Oncologist William Hrushesky of New York's Albany Medical College and his colleagues at the University of Minnesota assessed the impact of different treatment schedules of two commonly used anticancer drugs, adriamycin and cisplatin, in sixty women with advanced cancer of the ovaries. The women who took adriamycin a little earlier than their usual time of awakening and then took cisplatin twelve hours later—6 A.M. and 6 P.M., in most cases—fared much better than women who took the same doses of the same drugs in reverse order.

At one time of day, the drugs were more effective and less toxic. At the other time, they were more toxic and less effective. The reason for these differences apparently lies in the way the body handles the drugs. One adverse effect of cisplatin, for example, is kidney damage. The drug may cause less damage, Hrushesky said, if it is taken when it can be most rapidly excreted from the kidneys.

The women who received adriamycin in the morning and cisplatin in the evening had far fewer infections and instances of bleeding than those on the opposite schedule. They needed fewer transfusions. Fewer needed to have their dosages lowered or treatment postponed because of nausea, vomiting, or other adverse reactions. Most important, half of the women who received adriamycin in the morning and cisplatin in the evening were still alive five years later. Only 11 percent of those on the opposite schedule survived that long.

Another group of women received adriamycin and cisplatin at different times of day with no consistent sequence or interval between dosages. This remains the standard way these drugs are given at most of the nation's leading medical centers. Tragically, all of these patients died within three years. The side effects were so severe that the women typically dropped out of treatment after only three months. By contrast, those who received the time-specified treatment completed at least eight courses.

In another study, Hrushesky and his colleagues assessed timed treatment with the same two drugs in sixteen patients whose bladder cancer had been found during surgery to have begun to spread. Within two years of the surgery, this type of cancer ordinarily continues to spread in more than 90 percent of the patients. The timed drug regimen appeared to delay, and possibly prevent, further spreading. Eleven of the sixteen patients showed no recurrence of their disease when doctors examined them one to five years later.

A Canadian study published in 1985 focused on 118 children with acute lymphoblastic leukemia. After their initial treatment produced a remission, the children were given their daily maintenance dose of chemotherapy at home by their parents. Because it was not known whether the time of treatment made any difference, the parents chose the hour that was most convenient. The parents were instructed, however, to give the drugs at the same time each day. Georges Rivard and his colleagues at the University of Montreal in Canada found that 80 percent of the children who regularly received their medication after 5 P.M. were free of their disease five years after treatment began. Only 40 percent of those treated before 10 A.M. were disease-free—a clear demonstration, the researchers said, that evening treatment is better.

Timed treatment, despite its potential benefits, is not yet practical in most instances for general use. Timed drug delivery schedules are highly labor-intensive. Today's treatment regimens for cancer increasingly involve multiple drugs.

"Where would we get the staff to treat forty or even four patients at exactly six A.M.?" one oncologist wondered.

New programmable, automatic delivery systems may make chronotherapy for cancer, as well as for other diseases, more feasible in the future, when more and more people will receive their treatment as outpatients. Wearable or implantable devices are already being used in some medical centers. About the size of a hockey puck, the devices have one or more reservoirs for drugs. Eventually, doctors hope to have closed-loop systems that monitor the body's use of the various drugs or hormones and automatically adjust the dosage according to a predetermined formula.

Drug treatment is not the only area in which chronobiology studies may benefit cancer patients. Some additional avenues are illustrated by research on breast cancer, a disease that annually strikes 182,000 American women and causes 46,000 deaths in the United States alone. Some recent findings indicate that:

When surgery for breast cancer is performed may determine its success. Premenopausal women with early-stage breast cancer who had breast-removal surgery in the middle of their menstrual cycle—that is, around the time of ovulation—experienced fewer and later recurrences, and they lived longer than those operated on around the time of menstruation, according to Hrushesky and his colleagues who studied forty-four women five to twelve years after surgery. The hormone estrogen, which is at its highest midcycle, may exert a protective effect, the researchers speculated. Findings from this study, if replicated by other scientists, may prompt radical changes in scheduling surgery for breast cancer or use of supplemental hormones in women undergoing surgery.

Recognition of seasonal rhythms in breast cancer may improve the likelihood of discovering it in an earlier, more treatable stage. The incidence of breast cancer in premenopausal women peaks in the spring, and declines in the fall. In those who have completed menopause, it follows the opposite course. In premenopausal women, estrogen receptor concentrations, which are used to guide drug selection, are lowest in the early spring and highest in the late fall. In postmenopausal women, this pattern is reversed. Breast cancer causes more deaths in the spring. This suggests the necessity for more screening programs in the spring and perhaps more aggressive treatment at that time. Seasonal breast cancer rhythms may be related to seasonal fluctuations in sexual and thyroid hormones. The same mechanism may underlie some cancers in men: prostate cancer shows a spring peak and fall decline, and a type of cancer of the testes called *seminoma*—most common in men under age thirty-five—surges in winter.

The incidence of breast cancer appears to be higher in women living in northern cities with low levels of winter daylight than in women living in sunnier climes. Frank Garland, Cedric Garland, and their colleagues at the University of California at San Diego, reached this conclusion after comparing breast cancer rates with the amount of solar radiation striking the ground in eighty-seven areas of the United States. Lack of light may disrupt hormonal rhythms. Like the discovery of high rates of sleep disorders in the blind, the finding underscores the importance of light in assuring the normal functioning of the body.

Skin temperature in healthy breasts follows a predictable daily, weekly, and monthly pattern. These rhythms change if cancer develops, even when tumors are too small to feel: normal monthly rhythms fade, while weekly rhythms become more prominent, according to Hugh Simpson of the Royal Infirmary in Glasgow and his colleagues. The researchers fabricated an experimental "chronobra" to monitor and record breast skin temperature automatically. They envision that the chronobra would eventually be worn by women with a strong family history of breast cancer, who would use it to detect abnormal temperature rhythms just as they now perform breast self-exams or receive mammograms. It would be an additional aid to diagnosis rather than a replacement for such measures.

Asthma

Asthma, a breathing disorder whose name comes from a Greek term meaning "panting," has a profoundly circadian pattern: without treatment, asthma attacks occur fifty to one hundred times more often during the customary hours for sleep than in the daytime. They occur most frequently at 7 A.M., and least frequently at 3 P.M. A British study of nearly eight thousand people with asthma found that even among those who regularly took their medicine, four out of ten still awakened at least once every night wheezing and coughing. More than four thousand Americans die each year as the direct result of an asthma attack; such deaths are most likely to occur between midnight and 8 A.M.

Recognition of this pattern is not new. John Floyer, a seventeenth-century British physician who suffered from asthma himself, theorized that wheezing occurred more often at night, "where nerves are filled with windy spirits." Asthma affects one out of every twenty Americans—more than fifteen million people. It is the single most frequent cause of hospital admissions in children in the United States, and it is the leading cause of absenteeism from American schools.

Asthma causes the bronchial tubes and other airways in the chest to narrow or clog up, interfering with the normal flow of air in and out of the lungs.

Wheezing accompanies each breath, making a rough, raspy, whistling sound as air is forced through the blocked or constricted passages. People often cough in an attempt to loosen the excess mucus that clogs the airways and lungs; they may gasp for air and, in extreme cases, suffocate.

Allergies are the main cause of asthma; the long list of possible offenders includes pollens, molds, flowers, trees, weeds, grasses, house dusts, grain dusts, feathers, animal hairs and dander, and chemicals. People may also develop asthma as a result of irritation from fumes, smoke, and other industrial substances; in some people, asthma starts after a cold or other virus infection. In extremely sensitive people, coughing, laughing, crying, exercising, and even exposure to cold air can set off an asthma attack. Stress, anxiety, and other intense emotions do not cause asthma, but they may worsen an attack.

Monthly and yearly biologic rhythms further complicate the lives of people with asthma: the attacks often appear in women in the week before and during menstruation, probably because of hormone fluctuations. Yearly cycles in hormone production interact with increased pollens in the air in the spring and fall to make asthma attacks more common in those seasons. The illness is more severe in the winter months, when other respiratory diseases proliferate. Not surprisingly, deaths from asthma occur more often in the winter, too, particularly in those over age sixty-five who are hit hardest by the disease.

Because of the obvious day/night differences in asthma attacks, doctors have sought to minimize nighttime exposure to pillows, blankets, pets in the bedroom, and other presumed irritants. Such tactics may lessen the severity of the illness, but will not change its time course.

The impact of posture during sleep is still under investigation; because most people go to bed at night, it is hard to separate sleep from circadian rhythms. Being in a horizontal position narrows the airways and permits fluids to pool in the lungs, but nighttime changes in airway size occur even in people who stay awake. Fewer asthma attacks occur in the first half of the night, when sleep is deepest, but it has been suggested that deep sleep may suppress an allergic response, while the lighter sleep during the second half of the night may permit it to emerge. People with asthma typically sleep more lightly than those without the disorder. Moreover, loss of sleep often makes their breathing worse.

In the early morning, when most asthma attacks occur, the airways are most constricted, and in the midafternoon, when attacks are less frequent, the airways are most relaxed and open. In people with asthma, normal daily variation in airway size is much exaggerated: their peak airflow over the day may vary by as much as 50 percent, while in people without asthma, it varies by only 8 percent. The more severe the asthma, the greater the daily variation.

Paying attention to the pattern of daily variation can help identify a worsening condition before a crisis occurs. That is why doctors advise many asthma patients to use a peak-flow meter—an asthma "thermometer"—to measure at home how fast air can be blown out of the lungs at different times of the day. Recognition of the morning dip in airway size has also prompted doctors to monitor early-morning breathing more closely, particularly in patients who are at increased risk of suffering a potentially life-threatening bronchospasm at that time.

"As more is learned about how asthma changes on a twenty-four-hour basis, treatment should start following suit, but it may still be a while before pharmaceutical companies start designing medications to meet the needs of this disease," according to Richard Martin of the National Jewish Center for Immunology and Respiratory Disease at the University of Colorado. He called present-day therapy "hit or miss."

In recent years, asthma has been viewed less as a disease in which airways shut down and more as a disease of inflammation. This shift in thinking has altered treatment somewhat, although here, as elsewhere, doctors' drug-prescribing habits have resisted change. Bronchodilators to open constricted airways, such as theophylline, were once the most commonly prescribed medications and are still widely used; these may be inhaled or taken by mouth. There are more than three dozen types of bronchodilators available, and they are generally prescribed in equal daily doses. Inhaled corticosteroid drugs to suppress inflammation are becoming the mainstay of treatment; they are also generally prescribed in equal daily doses. Asthma's nighttime pattern has prompted some physicians to prescribe unequal doses of asthma drugs, two thirds of the drug to be taken at night and one third, in the morning, for example. Unfortunately, there are no studies of the drugs' safety and efficacy when used this way. The difference between the dosage that is safe and one that is not may be quite close, so identifying the times when larger doses could be given safely, or when smaller doses would work as well, would be an important advance.

Chronobiologically designed drugs also may enable people with asthma to rest easier soon. A once-a-day antiasthmatic drug available in Europe, but not yet in the United States, is specifically designed to give its peak effect between 2 and 6 A.M. to prevent nighttime wheezing.

Diabetes

Diabetes mellitus, or sugar diabetes, interferes with the way the body uses glucose, the simple form of sugar that provides energy. In healthy people, the hormone insulin, produced by the pancreas, enables body tissues to take up glucose from the blood. Insulin secretion follows a predictable daily cycle, with the peak secretion around 3 P.M. and the low around 3 A.M., with increases in response to the ingestion of food.

People with diabetes produce insufficient or no insulin, and too much glucose circulates in their blood, a condition called *hyperglycemia.* The body's efforts to remove the excess glucose—a homeostatic mechanism—causes both copious urination and excessive thirst.

For those who require insulin, it is tricky to adjust the dosage to mimic what the body ordinarily does automatically and precisely. Taking more insulin than needed will make blood glucose fall too low, producing *hypoglycemia,* which is accompanied by such symptoms as sweating, dizziness, and faintness. Tight control is important: results published in 1993 from a major national study that tracked more than 1,400 people with diabetes for more than six years indicated that the better the control of blood glucose, the fewer the complications the patients experienced.

There are two principal types of diabetes mellitus: insulin-dependent, or type I, the more severe form, usually appears in childhood or early adulthood; without insulin injections one or more times a day to counter hyperglycemia, people with type I diabetes would die. Non–insulin-dependent, or type II diabetes, usually develops in people over age forty and frequently can be controlled by diet, maintaining a normal body weight, and, sometimes, oral medications.

Glucose circulates in the blood for six to eight hours. That is why a continuous uniform dose of insulin, now possible with a wearable or implanted insulin pump, is not the complete answer for people with diabetes. Ideally, treatment would imitate the normal circadian pattern.

Some of the six million Americans with insulin-dependent diabetes experience an abrupt rise in the level of glucose in their blood, or in their need for insulin, or both, between 5 and 9 A.M. In some, the need for insulin may become six times greater than it was earlier in the night. Previously, those who experienced this so-called "dawn phenomenon" might have been advised to compensate for the change by increasing their morning insulin dose. Unfortunately, the added insulin sometimes had a rebound effect, causing blood sugar levels to fall too low later. Many doctors now recommend that the insulin dose be increased the night before in order to prevent or minimize a large early-morning change.

Knowing the daily insulin cycle enables doctors to tell their diabetic patients the best time to engage in exercise and other activities. Exercising in the early morning before breakfast when blood sugar is normally low, for example, carries less risk of hypoglycemia than late-afternoon exercise.

A better understanding of daily body rhythms has also given people with diabetes more employment options. Until recently, rotating shift work was regarded as impractical for most people with diabetes: the frequent changes in hours of work and sleep as well as mealtimes all complicate blood sugar control. It is still true that persons with diabetes usually find it easier to work permanent shifts with consistent physical demands that do not require constant tinkering with carbohydrate intake and insulin use. But "people with diabetes should not be systematically excluded from shift work," according to Gary Richardson of Harvard University. "With reasonable, intuitive adjustments of scheduling," he asserted, "the majority of people with diabetes can work on any shift." Diabetes is only one of many diseases affected significantly by work-schedule changes; for more information, see Chapter 10.

Arthritis

Biologic clocks determine the time of day that pain appears in the two most common forms of arthritis, or inflammatory joint disease. In osteoarthritis, also called *degenerative arthritis*, pain and stiffness are typically worse late in the day, reflecting wear and tear from daily activities. In rheumatoid arthritis, the pain is generally worse at the start of the day. Once attributed to immobility at night, this pattern now is recognized as the body's reaction to both immobility and the nighttime absence of cortisol, an inflammation fighter; after the normal 7 A.M. surge in cortisol secretion, pain lessens. Some thirty-seven million Americans, children as well as adults, have some form of arthritis.

People with rheumatoid arthritis who visit their doctors in the morning may appear to be in worse shape than they really are. In the afternoon, they may seem better than is actually the case. Physicians who see patients at different times of day from one visit to the next may have trouble telling how well treatment is working. Many doctors now ask patients with rheumatoid arthritis to measure grip strength and finger-joint swelling at home at different times of the day to help chart the course of their illness and the effectiveness of treatment.

Fatigue is a common complaint in rheumatoid arthritis, and this symptom has long been thought to be inseparable from the illness. Mark Mahowald and his colleagues at the Minnesota Regional Sleep Disorders Center in Minneapolis found it may simply be a symptom of poor sleep. All sixteen of the patients they studied moved their arms and legs excessively, and awakened frequently during the night. This is actually good news because it is much easier to treat the sleep problem than a vague ailment like fatigue.

Synchronizing arthritis medications with circadian rhythms may permit smaller or less frequent doses, thus reducing side effects. For rheumatoid arthritis, in which the joints are attacked by the body's own immune system, doctors commonly prescribe corticosteroid drugs to suppress the inflammatory response. These drugs work best and have the fewest negative effects when taken in the morning in sync with normal cortisol secretion. There is some evidence that cortisol-secretion rhythms are disturbed in people with rheumatoid arthritis, and that giving cortisol-based drugs at the right time can help normalize the pattern.

People with osteoarthritis who have nighttime or early-morning pain may get more benefit from pain-relievers in the evening. Those whose pain is greater in the afternoon or evening may find it more helpful to take pain-relievers in the morning or midday. Alain Reinberg and Francis Levi at the Rothschild Foundation in Paris studied five hundred patients with osteoarthritis of the hip or knee who took a commonly used pain-reliever, indomethacin, in a sustained-release form once a day at 8 A.M, noon, or 8 P.M. At different times of day, patients suffered only one quarter of the side effects, while gaining twice the pain relief. Patients differed in whether their best time was morning, noon, or night, a finding that attests to the need to tailor drug treatment to the individual.

Headaches

Some headaches wake people from sleep so frequently that they have been called alarm-clock headaches. These include *migraines*, typically worse in the morning, and *cluster headaches*, so-named because they come in groups. Both involve an abnormal stretching or dilation of the blood vessels. One study showed that three fourths of cluster headaches occurred between 9 P.M. and 10 A.M., with the highest frequency between 4 and 10 A.M. This points to a relationship between the headaches and REM sleep, a time of considerable variability in blood vessel size. Sleep laboratory studies have confirmed that cluster headaches often start in REM sleep or immediately after it.

Migraines occur mostly in women, and cluster headaches, in men. Some people have bouts of cluster headaches in the spring and fall, suggesting a link with seasonal hormone changes or changes in light. The typical female sufferer has 12 to 15 migraines a year, most around the time of menstruation, when estrogen levels are lowest. In one study, one third of the 512 migraines experienced by fifty-two women appeared four days before a menstrual period, and one third, while a period was under way. Shifts in hormone levels may themselves be a factor: birth control pills make migraines worse in some women, better in others. During pregnancy, some women's migraines disappear, but other women experience them for the first time. Migraines usually appear before age thirty-five, but some women develop them only after menopause.

Migraines cause dull pain, typically worse on one side of the head. Some people become very sensitive to light and experience nausea. Migraines usually last from two to six hours.

Cluster headaches, said to be the most excruciating of all headaches, may cause such intense pain—sometimes likened to being stabbed with a burning poker—that sufferers cannot lie down. Instead, they feel compelled to pace the floor or rock back and forth. Some bang their head against the wall. Cluster headaches begin as a pain around one eye that eventually spreads to that side of the face. They often cause reddening and tearing of the eye and a stuffed nostril on the painful side, and typically last about thirty minutes. Their periodicity is a unique feature that implicates an underlying disorder in the biologic clock: cluster headaches typically recur once or twice a day at the same times each day for weeks or months; then they cease for months, even years, before starting up again. Their predictability can be employed in treatment: taking the medication ergotamine a few hours before a headache is expected may lessen the intensity of the pain.

Fatigue, stress, and weather changes may trigger both migraine and cluster headaches. Glaring or flickering lights, and certain foods, including red wine and chocolate, have been implicated in migraines, while alcohol and a change in sleep schedule, particularly taking an afternoon nap, often induce cluster headaches.

Another type of *morning headache*, not necessarily violent enough to disrupt sleep but typically noticed on awakening, may stem from sleep apnea, a disorder in which breathing stops repeatedly during sleep. People with sleep apnea develop headaches because insufficient oxygen reaches the brain during the night. Sleep apnea occurs most frequently in overweight middle-aged men who snore raucously, the telltale indicator of impaired breathing.

The most common type of headache is aptly called a *tension headache*. Its name comes from the associated muscle tension, although emotional stress is a common cause. Here, muscles in the neck, face, and scalp tighten, producing a sensation of pressure being applied to the head or neck. Patients liken it to feeling their head trapped in a vise.

Some tension headaches recur at predictable times of the day or week, demonstrating how clock and calendar times can be transformed into biologic time. One teacher developed tension headaches at 11 A.M., just before she had to face a classroom full of unruly students. A salesman's headaches started at 5 P.M., when he reported the day's efforts to his manager. The stress of delivering a sermon causes a headache in clergymen so often that it has been dubbed a "preacher's headache." People who have unrewarding jobs often experience headaches on Saturday morning; during the week, they may keep their distress in check, but on Saturday, when they are freed from having to go to work, that is no longer possible. Additionally, they may sleep late, and schedule changes themselves may unleash tension headaches. Physical factors also can bring on muscle-contraction headaches at particular times of the day. Gripping the steering wheel of a car during rush-hour traffic, and hunching over books in an evening study session, are common examples.

The National Institute of Neurological and Communicative Disorders and Stroke reports that forty million Americans suffer chronic headaches. Seeking relief from severe, even disabling pain, they spend about $400 million annually on aspirin and other over-the-counter painkillers. Migraine sufferers alone lose more than sixty-four million workdays.

Better understanding of the rhythmicity of headaches may eventually lead to treatment that can forestall them. Recent studies, for example, have shown that one aspirin tablet (325 mg) every other day can help stave off migraines. Doctors' present arsenal includes several drugs designed to create a climate in the body that is less receptive to headaches. Many of these drugs alter the activity of serotonin, a brain chemical believed to play a role in the regulation of sleep and other biologic rhythms. Among these drugs are antidepressants that change the way brain cells communicate, and beta-blockers and anti-inflammatory drugs that relax blood vessels.

Epilepsy

Epilepsy, a disorder involving recurrent seizures, is also highly rhythmic in nature. Nearly 80 percent of people with epilepsy suffer seizures predominantly during sleep or on arousal from sleep. One study found that seizures occur most often between 7 and 8 A.M. Thus, shifting sleep periods around will change the time seizures occur.

The role that sleep processes play in triggering seizures remains unclear. People with epilepsy toss and turn and awaken frequently, perhaps compounding existing sleep abnormalities. Sleep deprivation so reliably intensifies seizure activity that it is used to elicit symptoms when epilepsy is suspected.

While anticonvulsant drugs appear to reduce both sleep problems and seizures, tailoring the drug to the individual patient is often more difficult and time consuming than for many other illnesses. According to the National Institute of Neurological and Communicative Disorders and Stroke, "sometimes the problem is one of dosage, or of finding the right combination of drugs administered in the right proportion at the right time of day."

Perhaps one quarter to one half of women with epilepsy find that their illness worsens during menstruation. In one study of 23,000 seizures that occurred in fifty institutionalized women over twenty-five years, researchers found that episodes occurred twice as often on the day a woman's period started as seven days earlier. This pattern persisted regardless of the regularity of the cycle or its length. Researchers suspect that progesterone may play a role in this pattern, perhaps by disrupting sleep.

Peptic Ulcers

Peptic ulcer pain strikes most often during the day between meals and between midnight and 2 A.M. Peptic ulcers affect only those areas of the gastrointestinal tract that are bathed by digestive juice: the lining of the stomach and the duodenum, which is the first part of the small intestine. Digestive juice, which helps break down food, contains hydrochloric acid and the enzyme pepsin—thus accounting for the use of the word "peptic."

Normally, the stomach secretes just the amount of digestive juice needed, when needed: secretion ordinarily peaks at midday and falls during sleep. When digestive juice appears at the wrong time, in the absence of food, it may erode the lining of the gastrointestinal tract, causing sores that produce a gnawing or burning sensation in the abdomen between the navel and lower end of the breastbone. Peptic ulcers affect one in ten Americans at least once in a lifetime.

Long-term use of aspirin and alcohol, as well as stress, may disrupt the normal secretion patterns. Frequent changes in mealtimes may play a role, too, as is suggested by the fact that ulcers occur more frequently in shift workers than in the general population.

Theories about the genesis of ulcers have undergone a dramatic change since research in the 1980s implicated a chronic bacterial infection of the stomach lining as a causative factor in ulcer formation. The bacteria are thought to neutralize stomach acid, providing them with a more hospitable place in which to live. Antiulcer drugs such as cimetidine (Tagamet) and ranitidine (Zantac) have been widely prescribed to heal ulcers. Antibacterial drugs and bismuth (the active ingredient in Pepto-Bismol), which coats the stomach, have been found to cure ulcers and prevent many recurrences by wiping out the offending bacteria.

One exception has been in people who take more than twelve aspirin a week—for arthritis, for example—who run an increased risk of suffering a peptic ulcer in the lining of the stomach. But circadian studies in animals suggest that *when* the aspirin is taken may either double the risk or halve it. In these studies, aspirin taken early in the morning caused little gastric erosion, while aspirin taken late in the evening caused considerable damage.

Leg Movement Disorders

Two related disorders that affect the legs—restless legs, which occur during wakefulness, and periodic limb movements (PLMs), which occur in sleep—have been found to have previously unsuspected daily rhythms.

In restless legs, the symptoms are disagreeable deep, creeping, crawling sensations in the calves and often the thighs, typically likened to electric shocks. The sensations usually occur when a person is sitting or lying down, prompting the urge to move the legs around and giving the disorder its name. The symptoms understandably often make it hard to fall asleep. Richard Allen and his colleagues at Johns Hopkins University asked patients with restless legs to stay in bed for thirty minutes before and after a night's sleep. They found that the subjects' legs moved two to four times as often at night as in the morning.

Nearly everyone who suffers from restless legs during the day also has PLMs at night, a disorder in which the feet, toes, and sometimes the knees, hips, and legs move abruptly and jerkily. Although these movements last only a second or two, they recur almost immediately, often hundreds of times during the night, causing extraordinarily restless sleep. Some people even wear out their sheets. The existence of an underlying disorder of the biologic clock is suggested by the fact that people with PLMs also experience unusual periodic fluctuations at twenty- to forty-second intervals in other bodily functions, including blood pressure, heart rate, and pupil size. William Culpepper of Bowling Green State University in Bowling Green, Ohio, and his colleagues studied seventy-two patients with PLMs. In these studies, two distinct patterns appeared: in one, the PLMs occurred mostly in the early part of sleep, and in the second, they were evenly distributed across the night.

Restless legs and PLMs are not rare; about one in ten people who complain of insomnia, and one in ten who experience excessive daytime sleepiness, has one or both of these problems. These two disorders become more common as people grow older. Several types of medication may be helpful in helping to alleviate symptoms: anti-Parkinson's drugs (although there is no relationship's between these disorders and Parkinson's disease), benzodiazepine sleeping pills, and opiates. Knowledge of the time course of these disorders has enabled doctors to adapt medication schedules to individual patients' needs. Some, for example, need medicine only at night or for only part of the night.

Weight Control

When one eats may be as important as *what* one eats in determining whether weight is maintained, gained, or lost. The now-classic studies in weight control, conducted in the mid-1970s by Halberg and his colleagues at the University of Minnesota, showed that people who ate nothing but the same 2,000-calorie meal once a day for a week lost weight when they ate only breakfast, but gained weight when they ate only supper. In these studies, breakfast-only meant that people ate the meal within an hour of awakening, and supper-only meant that they fasted at least twelve hours after awakening. Apparently, the body is more likely to burn fuel that is ingested in the morning and to store it in the evening. This information may benefit people trying to lose or gain weight; it also has far-reaching public health implications for the design of food relief programs for starving populations, suggesting that limited resources may go further if consumed late in the day.

The same researchers also showed that subjects on a breakfast-only regimen consumed fewer calories when they chose their own foods instead of eating those selected by the researchers. Surprisingly, they lost less weight on the self-selected lower-calorie diet than they did on the 2,000-calorie regimen. Eating foods one dislikes, the researchers concluded, may be a speedier way to lose weight than eating those one likes. Or the body may expend more calories in burning some foods—for example, vegetables and others high in fiber—than it does foods high in fat, such as ice cream. This notion challenges the nutritional axiom that "a calorie is a calorie."

There are circannual rhythms in body weight: humans act somewhat like hibernating animals in that they tend to put on pounds in the fall and shed them in the spring. This tendency is much exaggerated in people with winter depression, who often gain twenty pounds or more in two or three months. Many have more extensive summer wardrobes, reflecting more active summer lives. One man suffering from winter depression reported that his winter pants were two sizes larger than his summer pants. (For more on this disorder, see Chapter 8.)

MEDICINE IN THE TWENTY-FIRST CENTURY

Chronobiology research may transform the practice of medicine. A chronobiologic profile may become a standard part of a person's medical history—a map of the roads one's body travels from morning to night, day to day, winter to summer. Periodic chrono checkups may help to identify illnesses in their earliest, most treatable stages, and patients may participate actively in their own care by keeping self-measurements. This can be done today, as was demonstrated in a high school science class project in which students took their own temperature and blood pressure, and rated their moods six to nine times a day for several days.

The difficulty of acquiring time-of-day data has been until recently a major obstacle to the widespread application of circadian principles to medical practice. Since taking numerous samples of blood and urine is often impractical, chronobiologists have determined many time-of-day norms. Some labs already provide results with a time-of-day correction factor. The recent development of portable, ambulatory monitors to collect data, and of computer programs to analyze the information, has made around-the-clock assessments easier.

Computers will one day help determine the best timing and optimal dose of drugs. For some illnesses, portable, programmable drug pumps, implanted or worn on the body, will dispense drugs, monitor the body's use of them, and adjust the dosage as necessary. Ultra–short-acting drugs, targeted to specific phases of daily cycles, may be developed. Doctors' offices and outpatient treatment centers may stay open more frequently in the evening and at night.

More chronobiologic ways to deliver drugs may include long-term pulsatile polymer controlled-release systems. Already in existence are skin patches that are used to treat angina pectoris, for instance, and under-the-skin implants to prevent conception; these delivery systems release the drugs at decreasing, or constant, rates. By contrast, polymers can be made with many different physical properties to alter the rate of drug release. Hormones, for instance, could be delivered in a circadian pattern.

Drug testing will change, too. To date little heed has been paid to chronobiologic principles. Drug evaluation studies have traditionally employed a fixed dose; moreover, the rodents typically used are nocturnal animals, although researchers and patients generally are not. Chronobiologists question the logic of predicting human responses on the basis of tests in animals awakened from rest. Test results indicating that many substances are carcinogens, some suggest, may show merely the failure to consider biologic time.

The Food and Drug Administration (FDA) currently does not require manufacturers applying for approval of a new drug to conduct animal or human studies showing the influence of the drug's time of administration on its effectiveness or toxicity. Timed trials would be extremely costly, but money is not the only barrier. "You cannot make a requirement that would fit all drugs in all situations," John Harter, director of the FDA's pilot drugs evaluation staff, asserted. "Humans follow social schedules that are vastly different from those of animals. If you tried to study animals that were going to sleep and waking at different times from day to day, what event would you use to coordinate the effects of the medicine you want to give? What is the effect of meal timing, of varying meal timing, of skipping a meal?"

Although chronobiology is not yet a household word, a growing body of studies suggests that physicians—and patients—are beginning to think circadian. The sooner, the better, chronobiologists say. Factoring body clocks into treatment decisions, according to chrono-oncologist Hrushesky, "provides a golden opportunity to use what few imprecise chemical weapons we have a little more effectively."

CHAPTER 8

TREATING DEPRESSION

"I AM LIVING THROUGH THE NADIR OF MY LIFE," Michel Siffre, a French chronobiologist, wrote in his journal during the 205 days he lived by himself in Midnight Cave near Del Rio, Texas, in 1972, in one of the first, and still among the longest, time-isolation studies. "When you find yourself alone, isolated in a world totally without time, face-to-face with yourself," Siffre added, "all the masks that you hide behind—those to preserve your own illusions, those that project them before others—finally fall, sometimes brutally."

In 1980, reporter Dava Sobel chronicled for *The New York Times* her twenty-five days in the first chronobiology laboratory in the United States at Montefiore Hospital in The Bronx. "This laboratory is no desert island retreat," she wrote. "I have rarely felt more confined both physically and mentally. I cannot remember being this dependent and insecure in years." Ten years later, Sobel said that no adverse effects on her mood had lasted beyond her stay in the laboratory.

In 1988, in a study directed by Siffre, Véronique Le Guen, a thirty-two-year-old Frenchwoman, spent 111 days in a cave in southern France, the longest any woman had lived in time isolation. Le Guen came to believe that Siffre was torturing her. A sketch she drew to accompany a series of articles written for the magazine *le Figaro* while the study was in progress showed him as the author of books on sadomasochism. A year after the study ended, Le Guen died from an overdose of barbiturates; Paris police ruled her death a suicide.

Depression of this magnitude is extremely rare in participants in time-isolation studies. Daily ups and downs in moods are, however, both normal and predictable—in the laboratory and in everyday life. Moreover, people often become transiently irritable and "blue" when subjected to changes in their usual patterns of waking and sleeping, after jet travel or shift work schedule changes, for example. These experiences suggest that disturbed body clocks may cause disturbed moods, and perhaps tumble some into severe depression.

Time-isolation studies, of course, involve many stresses beyond those of time isolation itself: among these are separation from family and friends, the demands of the many tests that must be performed, and even the lack of exposure to daylight. The relative influence of each of these factors on negative moods remains to be clarified. Further, some people have reported that time isolation enhanced their sense of well-being. Both experiences are tantalizing to investigators, who have begun to take a new look at depression and other psychiatric disorders in a rhythmic context.

They have found that a large variety of rhythms, most notably circadian rhythms, often are askew in people suffering from depression. These include the rhythms of sleep, body temperature, and hormone release. Seasonal rhythms have also come under scrutiny. T. S. Eliot was on target in his assertion that "April is the cruelest month," for suicides peak in the spring—usually in May, however, not April. Admissions to mental hospitals for depression also soar in the spring. There is another, although smaller, surge in suicides and hospital admissions in the fall. What could account for such synchrony? The explanation presumably lies in some factor in the environment to which everyone is exposed but that triggers mental distress only in some. A likely contender: the length of daylight.

Cyclic patterns show that depression may originate from within, and is not simply, or not entirely, a reaction to adverse life circumstances. People who have one episode of depression usually have another, often many others. In the past century, physicians and patients alike commonly viewed these episodes as discrete and unconnected, more like hurricanes blown in from the tropics than the volcanic eruptions they more closely resemble. This is quite curious, given that "lunacy," the term once used for some forms of depression, comes from the Latin *luna*, meaning "moon," a language root that implies recognition of the periodic return of symptoms. "Because ours is a mobile society," Frederick Goodwin, director of the National Institute of Mental Health, has asserted, "depression's high recurrence rate has only recently become appreciated." However belated, the present-day rediscovery of its cyclic course has shifted attention from merely treating each acute episode as it comes along, to trying to prevent future ones as well.

While psychiatrists, psychologists, and other mental health professionals of every stripe, from the biochemical to the behavioral, are trying to solve depression's mysteries, chronobiologists are bringing something new to the table: tantalizing clues to heretofore unsuspected causes of the disorder, a better understanding of how psychoactive drugs work, and the development of treatments other than drugs, or that may make drugs work better, including manipulations of sleep and of light.

Improvements in treatment are sorely needed. Depression is becoming more common, particularly in people born since the mid-1940s, according to large comparable epidemiologic studies conducted in the United States, Sweden, Germany, Canada, and New Zealand. A larger percentage of the population was admitted to hospitals for depression in the thirty years since 1950 than in the thirty years before 1950. Depression is also showing up earlier in life than it did prior to World War II. Suicide rates in adolescents and young adults have tripled since midcentury: in the United States in 1951, the suicide rate for those aged fifteen to twenty-four was about 4 per 100,000, while by 1990, that rate had jumped to 13 per 100,000. Alcoholism and drug abuse have also become more prevalent. A better understanding of biologic clocks may hold the key to why these grim changes have occurred and, even more important, to what can be done to stem them.

THREE COMMON TYPES OF DEPRESSION AND THEIR SYMPTOMS

Major Depression

John Lennon reportedly was so depressed that he did not leave his bedroom for three years. Winston Churchill called depression a "black dog" that pursued him. Abraham Lincoln suffered so fiercely from black moods that he once wrote, "I am now the most miserable man living." Depression figures in the majority of suicides; among them were those of Virginia Woolf, Sylvia Plath, Ernest Hemingway, and Abbie Hoffman.

About ten million Americans experience depression in any given year, not simply the blues that weigh everyone down now and then in times of loss or disappointment, but a far more intense and unremitting state of melancholy. One in four women, and one in eight men, will suffer during their lifetime at least one bout of depression that lasts two weeks or longer, and is severe enough to interfere with work and family life. This type of depression, known as "major" or "clinical" depression, is so pervasive that it has been called the common cold of mental illness. Taking too long to fall asleep, and waking too early in the morning, two major disruptions of the sleep/wake cycle, are virtual hallmarks of this disorder.

In chronicling his own illness in an article for *The New York Times* (later expanded into his book *Darkness Visible*), William Styron addressed depression's often insidious course: "What had begun that summer as an off-and-on malaise and a vague, spooky restlessness," he wrote, "had gained gradual momentum until my nights were without sleep and my days were pervaded by a gray drizzle of unrelenting horror."

Styron's description of his experience is more graphic than that in most textbooks. "In depression, a kind of biochemical meltdown, it is the brain as well as the mind that becomes ill—as ill as any other besieged organ," he asserted. "The sick brain plays evil tricks on its inhabiting spirit. Slowly overpowered by the struggle, the intellect blurs into stupidity. All capacity for pleasure disappears, and despair maintains a merciless daily drumming. The smallest commonplace of domestic life, so amiable to the healthy mind, lacerates like a blade.

"Thus mysteriously, in ways difficult to accept by those who have never suffered it, depression comes to resemble physical anguish," he continued. "Such anguish can become every bit as excruciating as the pain of a fractured limb, or migraine or heart attack."

The common view of depression as a disease mainly of people in their forties, fifties, and older has not caught up with reality. Depression may afflict people of any age, even young children. A six-year-old in Florida committed suicide in 1993 by throwing herself in front of a train. A vulnerability to depression is inherited, with the risk about two to three times higher if one's parents, siblings, or other immediate family members also have experienced it.

Episodes of major depression last on average about nine months if untreated. While most people return fully to normal functioning, recent surveys show that four out of five will have recurring episodes, increasing in frequency over time.

Manic-Depressive Disorder

People with manic-depressive disorder experience alternating episodes of mania (highs) and depression (lows); another name for this combination is "bipolar disorder." Depression without mania is called "unipolar disorder."

Like a jacket that reverses from blue to red, mania is depression turned inside out. Mania is characterized by multiple excesses: those in its throes may be euphoric, overconfident, and overly optimistic; they also may be restless and talkative, with racing thoughts and easy distractibility. Some become irritable or angry. Buying sprees, sexual indiscretions, violent acts, or foolish business investments are commonly part of a manic episode, sometimes with devastating consequences for personal relationships and careers. A recent survey of patients with manic-depressive disorder indicated that more than one third were fired during an episode of their illness. Even more grim is the finding that people with mania who do not receive treatment have a suicide rate fifteen to twenty times that of the general population.

About 1 percent of Americans—an equal number of men and women—suffers from manic-depressive disorder, which usually starts between ages twenty and forty. Like other types of depression, this disorder runs in families. About one in eight of the immediate family members of someone with manic-depressive disorder will also suffer from it.

Manic episodes, if not treated, last about six months. In more than half of the cases, episodes of depression immediately precede or follow those of mania, with periods of normal functioning serving somewhat like bookends on either side. Some people, however, experience only episodes of mania or hypomania (a less frenzied state), interspersed with periods of normal moods.

A person might have a manic episode once, and then not have another for twenty years, a recurrence rate that has impeded recognition of the disorder's cyclicity. On average, however, manic episodes typically recur closer and closer together, settling in after the fourth or fifth episode to an approximately once-a-year cycle. People who experience extremely frequent cycles—every month or two, for instance—may have one or more "double days" at the start of each manic phase, forty-eight-hour nonstop stretches in which they skip one night of sleep entirely. These episodes serve as one of the most dramatic illustrations of a basic malfunction in the biologic clock.

Seasonal Depression

In this illness, which doctors call seasonal affective (meaning "mood") disorder, aptly abbreviated SAD, periods of depression consistently crop up at a particular time of year, most commonly in either winter or summer. Although similar to major depression, the seasonal types of depression tend to be more mild. Those afflicted may work at a reduced capacity, for example, but they seldom need to leave their jobs or enter the hospital. Their sleep is disturbed, but they do not show the REM sleep changes characteristic of people with major depression.

The most common type of seasonal depression is winter SAD. In his autobiography, *Clear Pictures*, Reynolds Price described his cousin Emma's winter mood changes, known in her community as "the kinks." "Only the gravest errand could compel anyone to attempt to visit her from Thanksgiving till early spring, though she seems to have been underfoot in her home," Price wrote. "Anyone calling with business for her farmer-brother was likely to glimpse Emma's narrow back, escaping to safety.

"But when we looked out one morning in late March and saw her walking up the road to church, someone could be counted on to say 'Miss Emma's out of the kinks.' And so she would be—a fountain of intense, intelligent and witty discourse till the next gray December."

People with winter SAD report that when they are ill, they may sleep four or more hours longer than usual each day (although they still awaken unrefreshed). They also overeat, particularly sweets and starches, often gaining twenty pounds or more in two or three months. This behavior may be related to the seasonal cycle of serotonin, a brain chemical whose presence may induce positive moods and whose absence may trigger depression. Serotonin is lowest in winter and highest in summer. Consumption of carbohydrates raises serotonin levels and may thereby temporarily improve mood.

In women with SAD, oversleeping and carbohydrate craving often become even more extreme premenstrually. Edvard Munch's painting "Melancholia" captures the winter SAD experience: a plump woman in a dark, shapeless dress slumps torpidly in a chair. Through the window behind her, one sees a bleak winter landscape.

In other seasons, people afflicted with SAD usually display normal moods, although some are simply less depressed and a few become hypomanic, or even manic; in general, they also become more energetic and creative, need less sleep, and lose weight. In highly creative people with winter depression, summer productivity may be dazzling. George Frideric Handel composed *The Messiah* in only twenty-three days in the summer of 1741. Vincent van Gogh completed twenty-nine paintings in June 1888, nearly one a day.

Winter SAD is thought to be brought on by a lack of exposure to daylight. Not surprisingly, sufferers are concentrated in northern latitudes, where winter days are shortest. Frederick Cook, the physician who accompanied Robert Peary's 1894 Arctic expedition, described a "gloom of the Arctic night" producing depression. "Although we had a very good Christmas dinner and everything we could wish for in the way of food," he wrote, "we were all very blue." Folk names testify to the pervasiveness of winter SAD. Alaskans call it "Arctic hysteria" and "cabin fever," and Norwegians refer to it as *morketiden*, "the murky time." In Iceland one word says it all: *skammdegistunglyndi*, the "heavy mood of the short days."

In a recent study conducted by researchers at the National Institute of Mental Health (NIMH) fourteen hundred patients who visited doctors' offices in New Hampshire, New York, Maryland, and Florida completed questionnaires. One in ten New Hampshire residents reported experiencing SAD symptoms, but only one in one hundred in Florida had such complaints. About 6 percent of Marylanders and 8 percent of New Yorkers described SAD symptoms. Researchers in England, Switzerland, and Norway also found that symptoms varied according to latitude. The farther north, the worse the problem. An estimated 5 percent of the population of the United States suffers from winter depression so acutely as to require treatment.

Many more people may experience similar but milder symptoms, from transitory gloominess and lethargy on a dreary day to more persistent winter doldrums. New York researchers surveyed four hundred people selected at random from the Manhattan phonebook. Fifty percent reported fatigue that increased during the winter. Nearly that many said they also needed more sleep, gained weight, and became more depressed gradually through the late fall, with the worst time occurring between December and February. According to Michael Terman and his colleagues at the New York State Psychiatric Institute, two million people in the New York metropolitan area alone may experience seasonal slumps. By the same token, spring fever is also a common physiologic phenomenon as well as a psychologic one—an upturn in mood triggered by the longer days of springtime. (See a self-test, "How Seasonal Are You?" on page 140.)

Winter depression typically first appears when people are in their twenties and thirties. It affects four times as many women as men, a predominance larger than in most other psychiatric disorders. It may occur in children, too, often causing changes in behavior that are more obvious than changes in mood. Children may complain that parents or teachers are harsh or excessively demanding, for example. While often written off as a school phobia, SAD symptoms generally start in December, whereas an expressed dislike for school usually emerges early in the school year. More than one third of those with SAD report that a close relative also suffers from winter depression, and two thirds have a close relative with major depression. Sometimes SAD appears only after people vulnerable to it move to a more northerly climate or start to work in winter months in windowless surroundings. A telephone lineman, for example, noticed SAD symptoms only after he took an indoor job.

Cultural influences may contribute to the disorder's severity, and even whether SAD symptoms are perceived to be a disorder at all. A native Alaskan businesswoman, living and working in the city of Fairbanks, complained that SAD kept her from doing her job well. The woman's sisters and brothers, still in their native village, had the same symptoms but no complaints, according to

SELF-TEST: HOW SEASONAL ARE YOU?

To what degree do the following change with the seasons?

	No Change	Slight Change	Moderate Change	Marked Change	Extremely Marked Change
Sleep length	○	○	○	○	○
Social activity	○	○	○	○	○
Mood (overall feeling of well-being)	○	○	○	○	○
Weight	○	○	○	○	○
Appetite	○	○	○	○	○
Energy level	○	○	○	○	○

At what time of year do you . . . ?

	Jan	Feb	Mar	Apr	May	Jun	Jul	Aug	Sept	Oct	Nov	Dec
Feel best	○	○	○	○	○	○	○	○	○	○	○	○
Tend to gain most weight	○	○	○	○	○	○	○	○	○	○	○	○
Socialize most	○	○	○	○	○	○	○	○	○	○	○	○
Sleep least	○	○	○	○	○	○	○	○	○	○	○	○
Eat most	○	○	○	○	○	○	○	○	○	○	○	○
Lose most weight	○	○	○	○	○	○	○	○	○	○	○	○
Socialize least	○	○	○	○	○	○	○	○	○	○	○	○
Feel worst	○	○	○	○	○	○	○	○	○	○	○	○
Eat least	○	○	○	○	○	○	○	○	○	○	○	○
Sleep most	○	○	○	○	○	○	○	○	○	○	○	○

or

No particular months stand out as extreme on a regular basis.

Using the scale below, indicate how the following weather changes make you feel. (ONE CIRCLE ONLY FOR EACH QUESTION)

−3 = In very low spirits or markedly slowed down
−2 = Moderately low/slowed down
−1 = Mildly low/slowed down

SELF-TEST: HOW SEASONAL ARE YOU?, *Continued*

0 = No effect
+1 = Slightly improves your mood or energy level
+2 = Moderately improves your mood or energy level
+3 = Markedly improves your mood or energy level

	−3	−2	−1	0	+1	+2	+3		DON'T KNOW
A. Cold weather	○	○	○	○	○	○	○		○
B. Hot weather	○	○	○	○	○	○	○		○
C. Humid weather	○	○	○	○	○	○	○		○
D. Sunny days	○	○	○	○	○	○	○		○
E. Dry days	○	○	○	○	○	○	○	OR	○
F. Grey cloudy days	○	○	○	○	○	○	○		○
G. Long days	○	○	○	○	○	○	○		○
H. High pollen count	○	○	○	○	○	○	○		○
I. Foggy, smoggy days	○	○	○	○	○	○	○		○
J. Short days	○	○	○	○	○	○	○		○

By how much does your weight fluctuate over the course of the year?

______ 0–3 lbs.
______ 4–7 lbs.
______ 8–11 lbs.
______ 12–15 lbs.
______ 16–20 lbs.
______ Over 20 lbs.

How long do you sleep each season?

	HOURS SLEPT PER DAY
Winter (Dec. 21–Mar. 20)	1 2 3 4 5 6 7 8 9 10 11 12 13 14 15 16 17 18 18+
Spring (Mar. 21–June 20)	1 2 3 4 5 6 7 8 9 10 11 12 13 14 15 16 17 18 18+
Summer (June 21–Sept. 20)	1 2 3 4 5 6 7 8 9 10 11 12 13 14 15 16 17 18 18+
Fall (Sept. 21–Dec. 20)	1 2 3 4 5 6 7 8 9 10 11 12 13 14 15 16 17 18 18+

SELF-TEST: HOW SEASONAL ARE YOU?, *Continued*

How to interpret your answers: a seasonal pattern, if one exists, should be apparent. People who feel worse in December, January, or February have a winter seasonal pattern. Those who feel worse in July or August have a summer pattern. *(See pages 137–143.)*

(Excerpted, with permission, from the *Seasonal Pattern Assessment Questionnaire*, by Norman E. Rosenthal, Gary H. Bradt, and Thomas A. Wehr. From: Rosenthal, Norman, et al. "Seasonal Affective Disorder and Its Relevance for the Understanding and Treatment of Bulimia." In *Psychobiology of Bulimia*, eds. J. I. Hudson and H. G. Pope. Washington, D.C.: American Psychiatric Press, 1987, pp. 205–228.)

psychiatrist Carla Hellekson. "When people live by hunting and trapping," Hellekson observed, "they do not view their need to sleep longer in winter, or to store carbohydrates and gain weight as significant problems."

In the spring, when people with winter depression start to perk up, an estimated 1 percent of the population finds that moods start to sink. Those with summer depression suffer many symptoms that are the reverse of their winter counterparts: increased agitation, a decreased need for sleep and food, and weight loss. Those with summer SAD, however, are more likely to have suicidal thoughts.

Like winter SAD, summer SAD is related to latitude: it is more severe, and more frequent, in southerly regions, and it is rare in northerly climes. Researchers in Iceland were unable to identify a single case. The English poet John Milton may have had summer SAD. He told a friend that his summertime work "was never to his satisfaction," and that during the many years he spent writing *Paradise Lost*, he had "spent but half his time therein."

Excess light does not appear to be the culprit in summer depression. Rather, other environmental factors, most notably excessive heat and humidity, are thought to be at fault. A college professor who had lived in northern Europe and New England developed depression in the warm months only after he moved to Washington, D.C. When he went to New England for summer vacations, he felt better in two or three days. While visiting Europe during an unusual heat wave, he relapsed.

SYMPTOMS OF SAD

To qualify as SAD, the symptoms below must appear at a characteristic time of the year, and they also must disappear at a characteristic time of the year for at least two years.

- Difficulty in finishing usual tasks, including work assignments or household chores.
- Difficulty in getting to work on time.
- Problems in thinking and concentrating.
- Feelings of sadness, guilt, or pessimism; crying spells.
- Negative thoughts about oneself in one season but not at other times of the year.
- A need for several hours more sleep than usual; excessive fatigue.
- Difficulty in controlling food intake and weight.

CHRONOBIOLOGIC TREATMENTS FOR DEPRESSION

Sleep Therapy

That there is a link between depression and disruptions in the sleep/wake cycle was recognized in ancient times. The second century Greek physician Aretaeus noted that people who were "sad, dismayed, and sleepless," often "became thin with their agitation and loss of refreshing sleep."

Recent research has shown that the appearance of sleep problems may even signal that a major depressive episode is brewing. In one study, Daniel Ford and Douglas Kamerow of the NIMH asked people whether they got enough sleep, or felt excessively sleepy during the day. They asked the same people the same questions a year later. Those who had had either of these problems when first surveyed, and still had it a year later, proved more likely to have a severe depression or anxiety disorder than those whose sleep problems disappeared within the year, or who had no sleep complaints.

Beginning in the early 1970s, scientists used the sleep laboratory to study people with severe depression, finding many abnormalities, mainly in REM sleep. In these people, the first REM episode of the night started too soon, often within twenty minutes of falling asleep, David Kupfer and his colleagues at the University of Pittsburgh discovered. In normal sleepers, the first REM episode usually started about ninety minutes after sleep began. In general, Kupfer found, the earlier the first REM episode occurred, the more severe the depression.

Other REM sleep characteristics differ, too. In normal sleepers, the first REM episode lasts only about ten minutes, the second, about fifteen minutes, and each successive episode lasts even longer. In those suffering from depression, however, the first REM episode may be the longest, containing many more rapid eye movements than is normal. The resulting lengthy, busy dreams may be more emotion laden, complicated, bizarre, anxious, and full of frustration than dreams that start at a normal time, and may help to perpetuate the depression.

The various REM sleep abnormalities diminish but do not entirely disappear when depression lifts; they thus serve as scars of the illness. Moreover, relatives of people with depression, particularly those who have experienced depression themselves, often show the same aberrations. Kupfer and many others in the field see the REM changes as biologic markers—stigmata, some say—of a vulnerability to depression. While these markers may affirm the genetic nature of the illness, researchers hope that by detecting and studying people with such signs, ways can be discovered that will both uncover and treat depression earlier. A doctor who discovered that a depressed patient's REM sleep was occurring earlier and earlier in the night, for example, might want to increase the dose of antidepressant medication the patient was using or take other measures to forestall a depressive episode.

The disruption in the timing of REM sleep is only one of several important body rhythms that often run too early in people with depression; others include the body temperature cycle and the release of the hormone cortisol. The most active dreams ordinarily occur at the end of sleep, a rise in temperature induces wakefulness, and the level of cortisol increases with stress and promotes arousal. It is easy to see that if these events occur too early in the night, they might cause people to sleep poorly.

The connections between REM sleep and depression have already generated a new treatment: sleep therapy. This therapeutic approach arose from the observation that antidepressant drugs markedly reduced REM sleep and sometimes even wiped it out entirely. Researchers wondered whether the success of these drugs actually might lie in their ability to curb REM sleep. If so, they reasoned, there might be better ways to accomplish that task.

To pursue this goal, Gerald Vogel of the Georgia Mental Health Institute performed an ambitious experiment: in the mid-1970s, he monitored depressed patients sleeping in his laboratory, and awakened them every time brain waves and eye movements showed that they had entered the REM state. Because a brain deprived of REM sleep tries to recover it as soon as sleep resumes, Vogel sometimes had to rouse people thirty times a night.

Half of a group of hospitalized patients who participated in this strenuous regimen six nights a week for seven weeks got better. They did as well as a matched group of depressed patients who took antidepressant drugs, Vogel found, and they improved more than patients awakened during NREM sleep. Indeed, they were able to leave the hospital and resume their work and family lives. Two years later, most of the subjects remained well without needing antidepressant drugs, or further sleep treatments.

As a workaday treatment for depression, rigorous REM deprivation is impractical: it requires a sleep laboratory with a battery of electronic devices to record brain waves and eye-muscle activity, and a technician to monitor the printouts and dash in to wake people whenever they slip into REM sleep. But the success of the experiment provided the jumping-off point for further studies.

Looking for shortcuts, researchers decided to try keeping depressed patients awake all night. Although this approach obviously could not be continued indefinitely, they hoped it would provide a boost to sustain the patients for a few days or even weeks. Asking people who already slept poorly to give up a night's sleep is, on the face of it, quite a peculiar idea; perhaps the total loss of sleep would make their black moods worse. Convincing patients to volunteer took some doing.

Christian Gillin of the University of California at San Diego described a woman preoccupied with her own worthlessness to the point of being suicidal: "She usually went to bed around midnight. By 5 A.M. on the first night we kept her awake, she started feeling better. By 8 A.M. she was almost completely recovered. Previously, she'd been nearly immobile and barely talked, but now she acted and conversed normally.

"She continued to feel well throughout the day. At 3 P.M. we allowed her *one minute* of EEG-recorded sleep. She awakened abjectly depressed. She insisted that she had not slept. She accused us of injecting her with a drug."

This striking experience has occurred again and again. Scientists have now kept hundreds of depressed patients awake for all or certain parts of the night, for one night or for several nights. They have interrupted slumber to deprive patients of one specific state of sleep. They have also put patients to bed late, and awakened them early. And every one of these approaches, Gillin reports, has brought about improvement, at least temporarily, in one third to two thirds of the patients—rates that rival antidepressant drugs in effectiveness.

The success of such manipulations raised some intriguing questions: does sleep per se, or a certain amount or state of sleep, somehow bring about depression or make it worse? Is there a flaw in the timing of sleep or in the timing of a particular state of sleep? Do the roller-coaster changes in mood follow the changes in sleep or precede them?

It turns out that most switches into depression occur during sleep; that is, a person goes to sleep seemingly well, but awakens depressed. Researchers have observed that many depressed people start their days in gloom, but brighten up as the hours pass; one theory is that the depressant effect of the night's sleep may take some time to wear off. When their depression returns the next night, it is no wonder that if they awaken in the predawn hours they may feel as if they have opened the lid to a Pandora's box of morbid thoughts. This experience may have prompted F. Scott Fitzgerald, who himself suffered from depression, to write, "In a real dark night of the soul it is always three o'clock in the morning."

For a few patients undergoing sleep therapy, loss of sleep made the pendulum swing too far: it caused them not only to switch out of depression but into mania. Most frequently cycled between depression and mania, say, four or five times a year. This finding means that sleep deprivation must be used cautiously in anyone with a history of mania. It also implies that preventing disrupted sleep in such persons might prevent some manic episodes, and that their family members should be alerted that missed sleep might signal an impending episode.

The observation that sleep often had a depressant effect, while wakefulness boosted spirits, fostered the development of additional new ways to improve treatment for mood disorders. To diminish the depression-inducing effects of sleep, researchers tinkered with the amount and timing of sleep. They discovered that missing only part of a night's sleep was as effective in relieving depression as staying awake for the whole night, and that skipping the second half of the night was more beneficial than skipping the first half.

They also found that shifting sleep to an earlier time improved moods. Thomas Wehr and his colleagues at the NIMH asked depressed patients to go to bed at 5 P.M. and get up at 1 A.M., six hours earlier than usual. Their depressions all abated. The benefits lasted only one to three weeks, however, and a rerun did not give as good results as the first experiment. Shifting the hours in bed made REM sleep occur later in the sleep period. While this maneuver provided only a temporary fix, the fact that it worked at all was important; it suggested that the rhythm disturbance had caused the patients' depression and was not merely a consequence of it.

In other studies, scientists found that combining a change in sleep time with the use of antidepressant medication may make both treatments work better, and produce longer-lasting benefits than either treatment by itself. In one study, David Sack and his colleagues at the NIMH asked four depressed patients who had shown no signs of improvement while taking antidepressant drugs to go to bed several hours earlier than usual. After three days on the new schedule, all the patients got better. The patients continued taking their medication, while their sleep schedules were gradually shifted back to their customary times. Several months to a year later, they continued to maintain improvement.

While much remains to be done to refine sleep therapy, there are many good reasons for using it today. Since the benefits occur rapidly, it may reduce the risk of suicide, and shorten the two to three weeks a seriously depressed patient needs to stay in a hospital waiting for drugs to take effect—a major incentive in the face of upwardly arching medical-care costs.

As not everyone can tolerate antidepressant drugs, or is willing to take them, sleep therapy may enable some depressed patients to avoid drugs altogether. Women with premenstrual syndrome, a disorder characterized by multiple disturbances of biologic clocks (see Chapter 5), may do well by undergoing sleep therapy only during the week or so that their symptoms surface—a more attractive treatment schedule than taking medication all month long. Barbara Parry, of the University of California at San Diego, and Thomas Wehr found that eight of ten women with PMS who went one night without sleep during the premenstrual phase of their cycle reported that their depression had lifted, and improvement continued even after a night of recovery sleep. One woman said that she was still doing well six months later. Five of those who had improved with total sleep deprivation also did well after skipping sleep only in the latter half of the night.

Light Therapy

Every fall and winter for more than twenty-five years, Herbert Kern had plunged into depression. "My energy level sank," he recalled. "I felt anxious and fearful for no reason. My memory went sour. I avoided people. I put on a few pounds. My sex life went to pot. I fell into bed at 9 or 10 P.M., yet I had to force myself to get up in the morning." A research chemist living in Florham Park, New Jersey, Kern observed that his mood improved with the coming of spring. "The wheels of my mind started to spin again, and my work productivity soared," he said. "I needed only three or four hours of sleep. I also enjoyed socializing again."

Over the years, Kern had consulted several physicians; he had tried different antidepressant drugs, but found their side effects intolerable. Using a daily diary that he had kept for many years, he plotted the seasonal variation of the sun's intensity, concluding that it had a direct relationship to his highs and lows. In 1980, Kern, then sixty-two, took his hunch to psychiatrist Alfred Lewy at the NIMH, who had just reported a new technique for measuring the hormone melatonin in the blood. Lewy and his colleagues found that melatonin rises ten- to fiftyfold in the dark. Kern was looking for a light-responsive biochemical marker that might correlate with his seasonal mood swings. He wanted Lewy to measure his melatonin levels at different times of year.

Kern's request was fortuitously timed, for Lewy and his team had also discovered that melatonin secretion in humans could be turned off by exposure to daylight-equivalent artificial light, a finding that was to change Kern's life. Researchers had learned a few years earlier that a brief exposure to light could shift melatonin secretion in animals, but several studies had suggested that was not so in humans. In fact, as Lewy's team demonstrated, the lights used in the human studies had been too dim. By proving that light could alter the human melatonin rhythm, the NIMH work provided strong evidence that it might govern other human rhythms as well.

The NIMH team devised a novel treatment for Kern. They asked him to come to the NIMH campus in Bethesda, Maryland, in December when his depression was at its worst, and to sit facing a bank of bright fluorescent lights for three hours in the morning, and for another three hours in the early evening. The lights, which were five to ten times brighter than ordinary room lights, were equivalent to daylight just after sunrise. By lengthening Kern's day, the researchers in essence made spring come early. After just four days, his gloom lifted. "For the first time in many years," Kern recalled, "I truly enjoyed Christmas."

A newspaper article about Kern brought the researchers bags of mail from people similarly afflicted, many eager to participate in research studies. An article, "Seasonal Affective Disorder: A Description of the Syndrome and Preliminary Findings with Light Therapy," by NIMH researcher Norman Rosenthal, along with Lewy and others, appeared in the *Archives of General Psychiatry* in January 1984, and the rest, as they say, is history. If exposure to light reversed the symptoms of winter depression, the researchers reasoned, then lack of light must be causing the disorder.

Although the procedure Kern used so successfully has undergone some refinements, daily exposure to artificial bright light remains the mainstay of treatment for winter depression. Thousands of patients have had light therapy, and three out of five of them have gotten better. Researchers initially mimicked a spring day, giving light treatment before dawn and after dusk, but they soon discovered that they did not have to extend the day. Light given during normal daylight hours proved sufficient.

Timing of light exposure remains a matter of some debate, although there is general agreement among researchers that most patients derive the greatest benefit from light received between 6 A.M. and 8 A.M. Still, some patients do better with afternoon or evening exposure, which has prompted speculation about the nature of the underlying flaw in seasonal depression, and how light might correct it.

Lewy, now at the Oregon Health Sciences University in Portland, has posited that some people with SAD—the ones who have a hard time awakening in the morning—may have circadian rhythms that run too late in relation to the natural alternation of day and night. Bright light in the morning, he has suggested, may reset their body clocks to an earlier schedule. By contrast, those who awaken too early may have rhythms that run too early; bright light in the evening may shift their clocks later. Clinicians need to determine which type a patient is, Lewy asserts, before they prescribe light therapy.

Alternatively, SAD sufferers may have a hunger for light that they can satisfy at any time of day, Rosenthal has proposed. If light per se is energizing, how much of it they get may prove to be more crucial than when they get it. An unsatisfied need for light may explain why people with SAD often feel worse in January and February, when the days are longer, than in December, Rosenthal has said. In the first two months of the year, the days are usually more overcast. If correct, this theory would have enormous practical significance: patients could schedule treatment whenever it was convenient. Time may be an important factor in undergoing the treatment at all, since as much as two hours of daily exposure may be required for maximum effectiveness.

Initially, it was hypothesized that some flaw in melatonin secretion played a role in triggering winter depression, but numerous studies failed to confirm this. In one study, researchers gave melatonin during the daytime to patients who had improved with light therapy to see if it brought their depression back. That did not happen. In another study, patients took a drug called atenolol that, like light, stops melatonin secretion, but they did not get better. The length of melatonin secretion is now thought to be simply a marker that tells what time, or what season, it is on the body clock.

Interest has turned to the role of a deficiency of the nerve cell messenger serotonin in triggering SAD. Without serotonin, there would be no chemical reaction resulting in melatonin. The emergence of SAD in the winter coincides with serotonin's annual low, and SAD symptoms remit in the spring as serotonin levels rise. Antidepressant medications that increase the amount of serotonin, such as fluoxetine (Prozac), sertraline (Zoloft), and paroxetine (Paxil), all relieve SAD symptoms. For some SAD patients, a combination of one of these drugs and light therapy is the most effective treatment.

Much research has focused on determining what time of day exposure to light works best, how bright the lights need to be, and how long an exposure is required. Light intensity is measured in "lux." Artificial light of about 2,500 lux—the intensity level of early daylight—seems to be a threshold for bringing about improvement in SAD symptoms. By comparison, starlight is less than one lux, conventional room light is around 500 lux, and a reading light may provide 1,000 lux. All natural daylight, even if the day is overcast or it is raining, is brighter than ordinary indoor light. On a bright summer day, sunlight at noon measures about 100,000 lux.

Most SAD patients benefit from bright light exposure every day beginning in the fall, when their symptoms usually appear, and continuing through the winter months. Relief from symptoms occurs within three to five days (antidepressant drugs may take three to five weeks to effect improvement); likewise, if treatment is stopped during the winter, the symptoms return in about three to five days. Although the initial treatment for SAD called for two hours of daily exposure to 2,500 lux light, by 1993 many researchers had concluded that it was just as effective, and perhaps even more so, to increase the light intensity to 10,000 lux, the equivalent of daylight about forty minutes after sunrise, and to use the light in the morning for only thirty minutes. Battery-operated light visors and light devices that clip onto eyeglasses or caps also are being marketed, allowing patients greater freedom of movement.

The most widely used light source is a bank of fluorescent lights mounted on a metal reflector, commonly referred to as a "light box." Patients customarily sit about two to three feet away from the lights. They are permitted to read, eat meals, or watch television. They are instructed to gaze at the lights frequently, but not to stare at them. The early studies used lights that mimicked natural sunlight, but more recent studies suggest that full-spectrum light is unnecessary; the ultraviolet component, which theoretically may damage skin and eyes, can be screened out.

Light boxes may be purchased without a prescription for about $400. However, there are good reasons to seek professional advice: since other illnesses, including thyroid disorders and chronic viral infections, may masquerade as SAD, a physician or psychologist can determine if a self-diagnosis is correct, and then prescribe a specific light regimen and monitor its effectiveness.

Whether light exerts its antidepressant effect through the eyes or through the skin once was debated. There are many reasons to suspect that skin exposure might play a role: sunlight enables the skin to manufacture vitamin D (a misnomer; it is a hormone), needed for healthy bones and teeth; light shining on the skin remedies jaundice in newborn babies; the warmth of the sun on the skin feels good; and on sunny days, people report more optimistic moods.

To resolve this issue, Thomas Wehr and his colleagues at the NIMH compared the effects of skin and eye exposure to the same light source. They enlisted ten people with winter SAD to spend four hours each evening in front of bright lights for two weeks. During one week, the participants wore bathing suits and shielded their eyes with dark goggles; during the other, they wore clothing from head to toe, including face masks that left only their eyes uncovered. Given standard measures of depression before and afterward, seven of the ten participants improved with the eye treatment, but not with the skin treatment. One individual got better with the skin treatment, but not the eye treatment. One responded to both treatments, and one to neither. The study's bottom line: the eyes have it.

Given this finding, it is all the more surprising that light exposure that begins in the morning while people are still sleeping also may improve moods. Some light evidently comes through closed lids; moreover, the retina is most sensitive to light in the early-morning hours. A person who wished to live in sync with the natural world would sleep with the shades up in an east-facing room, so as to be awakened gradually by the rising sun; failing that, one could utilize a computer-assisted device to turn lights on gradually. Several studies have shown such devices to be particularly useful for people with winter depression who also have difficulty awakening in the morning.

Despite the apparent success of light treatment, there are many skeptics who view SAD as a medical fad and light treatment as a brilliant placebo. In psychiatry, where the motivation to get well is critical to the success of any therapy, the placebo—meaning, "I will please"—response to treatment is said to account for 20 to 40 percent of the effectiveness of psychiatric drugs, at least initially. In most studies of light therapy, bright lights have been compared with dim lights; while the former have brought about improvement in 60 percent of the patients on average, even the latter have yielded a 20 percent improvement rate.

Charmane Eastman of the Rush-Presbyterian-St. Luke's Medical Center in Chicago and Henry Lahmeyer of the University of Illinois Medical Center devised an ingenious experiment to determine if light could outperform a different type of placebo. They compared exposure to light with exposure to negative ions, another natural environmental component. Like light, negative ions are higher in the summer than the winter; some SAD sufferers claim that negative ions make them feel better. When surveyed in advance, patients thought the two treatments would be comparable. The researchers asked some subjects to sit close to the lights, others to a working negative ion generator, and still others to a deactivated negative ion generator. While even the sham device made some people better, light therapy produced the most improvement.

In another study of the placebo effect, P. Richter and his colleagues at the University Hospital in Groningen, the Netherlands, hypnotized volunteer SAD patients, and told them to imagine that they were sitting in front of bright lights. After three days of visualizing lights, the patients had ten days off, and then they were exposed to real lights for three days. The patients showed some improvement with the imaginary lights, but the real lights worked much better, and their benefits lasted longer. The fact that many patients continue to benefit from using light therapy year after year also argues against its being merely a placebo.

The side effects of light therapy are quite modest compared with those of antidepressant drugs, and they generally can be controlled by reducing exposure. Some people report headaches, eye strain, or jittery feelings. It is rare for hypomania or mania to appear after light therapy, and there is no indication that anyone has experienced any eye damage after being exposed to lights that, after all, are comparable to only low levels of ordinary daylight. Nonetheless, it is customary for clinicians to advise patients to obtain eye exams before beginning treatment.

Considering the usefulness of light treatment for winter depression, it is not surprising that researchers have sought other applications. Daniel Kripke and his colleagues at the Veterans Administration Hospital in San Diego used bright lights to treat men hospitalized for severe clinical depression. The men improved, but, according to severity measures, only about half as much as people with winter depression who had received the same treatment. The results were nowhere near as dramatic as those achieved by sleep deprivation.

Bright light may help people who have trouble sleeping or staying awake at the appropriate times of day. (See Chapter 6.) Light therapy provided no benefits, however, for women with bulimia nervosa, another disorder that is worse in winter and also involves food cravings and overeating. And while bright light reduced PMS symptoms, Barbara Parry found that the women improved just as much after being exposed to dim light. This is not surprising given that other studies have shown that even dim light may alter reproductive cycles.

Bright light has been long reputed to speed physical healing. Florence Nightingale was among the first to champion sunrooms in hospitals. More recently, geographer Roger Ulrich of the University of Delaware compared two groups of patients who had undergone a common type of gallbladder surgery. The patients whose rooms faced a stand of trees and who could see changes in daylight fared much better than those who looked out on only a dreary, brown brick wall. The former group needed fewer painkilling drugs, were judged by nurses to be improving faster, and went home sooner.

Cold Therapy

A woman who lived in Washington, D.C., had suffered from frequent crying spells and suicidal thoughts for twenty years, usually beginning in April and ending in September. She improved dramatically during a brief cold spell one June, just a few days before a scheduled consultation at the NIMH. When her depression returned, the NIMH researchers asked her to stay in her air-conditioned home for five days, and to take cold showers for fifteen minutes several times a day. On this routine, her depression lifted. Unfortunately, a few days after she returned to her normal schedule and ventured out into the summer heat, it returned.

She was among twelve people with summer depression who volunteered for a study at the NIMH in 1986. For five days, the volunteers received "cold treatment": they stayed in air-conditioned surroundings, and spent twenty minutes four times a day between two cooling blankets set at 40°F. During another five-day period, they received "dark treatment": they stayed indoors, wore dark goggles, and sat in total darkness for twenty minutes four times a day. They got better after both treatments. The next winter, when they were well, the twelve subjects were reexposed to summer conditions, once to bright light and at another time, to heat and humidity. The light did not bother them, but the heat and humidity made them feel worse, which implicates faulty thermoregulation as a core issue, according to Thomas Wehr, director of this work.

DEPRESSION AND THE AMERICAN WAY OF LIFE

Seasonal depression was first included in the *Diagnostic and Statistical Manual of Mental Disorders* of the American Psychiatric Association only in 1987. The increasing prevalence of this disorder, along with the big jump in major depression in the past half century, has prompted considerable academic head-scratching. What might have prompted these events? One may safely assume that people have not changed markedly in recent years; that makes one or more environmental influences the more likely culprits.

A round-up of the usual suspects includes greater geographic mobility, with its resulting loss of attachments; higher rates of divorce; more women in the workforce; and more widespread financial problems—all implicated in the development of depression by numerous studies. But in addition, some modern amenities—particularly artificial lighting, air-conditioning, and central heating—may expose people year-round to "insults" to their biologic clocks against which they have no natural protection, and thus provoke mood and behavior disturbances.

If even one of these factors—namely, artificial light that lacks the intensity of natural daylight—produces depression, then millions of people may be at risk. People born since World War II, living in industrialized societies, are unwitting participants in a wide-reaching experiment: they are the first people in history to spend most of their lives under artificial light—even when engaging in sports, such as tennis and jogging, that once brought them outdoors. However, people, like petunias, may need a certain amount of natural light to thrive.

Indoor lighting design emphasizes visual acuity, not biologic rhythms; humans can see perfectly well in light that is much lower in intensity than daylight. As a result, most urban-dwelling Americans may be living chronobiologically in the dark. According to Richard Wurtman of the Massachusetts Institute of Technology, people living in Boston who spend sixteen hours a day in a conventionally lighted indoor environment, are exposed to less visible light than if they spent a single hour outdoors each day.

Even in San Diego, one of the nation's sunniest cities, most inhabitants receive astonishingly little bright light exposure. In one study by Daniel Kripke and his colleagues at the University of California at San Diego, more than one hundred volunteers wore computerized light detectors on their wrists all day as they went about their normal activities. It was discovered that even in the summer, healthy young adults received an average of only ninety minutes of outdoor light a day, mostly in going to and from work.

Healthy elderly men in this study averaged seventy-five minutes of outdoor light a day, but elderly women, only twenty minutes—a fact that may help explain why elderly women have higher rates of sleep complaints and depression than elderly men, Kripke suggested. People with Alzheimer's disease, living at home, averaged only thirty minutes of outdoor light, and residents of nursing homes, only two minutes.

The impact of bright light on people with normal moods is just beginning to be investigated. A Swiss study found that people who received an hour of bright light every morning and evening for one week reported that their feelings of well-being improved. They did not, however, feel more energetic. Anna Wirz-Justice and her colleagues at the University of Basel found that the mood lift lasted as long as four weeks. A half hour of light exposure twice a day for one week had no effect.

When NIMH researchers tried a similar experiment, they found that many people who regarded their moods as normal did, in fact, report subtle mood changes in the winter, such as a mild loss of energy or diminished interest in socializing. These were the only ones to benefit from light exposure. People whose moods were the same year around derived no boost from receiving additional light. According to Siegfried Kasper and his colleagues, "the study suggests there are individuals who do not seek medical help but might nonetheless benefit from it." Indeed, for such people, vacations in sunny climes may serve as a self-administered therapy.

The need for bright light exposure has major implications for the design of homes, offices, schools, and other public places as well as for environments in which access to the outside is restricted, such as submarines and spacecraft. In the future, Michael Terman has suggested, "supplementary bright light will prove to be more analogous to air-conditioning and humidifying than to a medical procedure." It is no surprise that supermarkets, shopping malls, and fast-food restaurants are already dazzlingly bright. While there are few academic studies showing the mood-elevating and energy-boosting effects of bright light in such surroundings, the sound of the cash register must be quite persuasive.

PART III

BODY CLOCKS AND MODERN LIFE

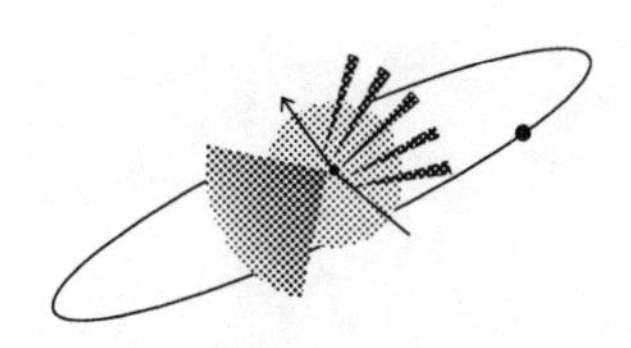

CHAPTER 9

COPING WITH JET LAG

IN JUNE 1955 IN PARIS, TWO GERMAN RESEARCHERS, Max Renner and Karl von Frisch, aided by French colleagues, installed a hive of bees in a room where light, temperature, and humidity were kept constant. Renner spent two weeks training the bees to expect sugar water to appear on a table in a corner of the room between 8:15 and 10:15 A.M., French summer time. Then he and the bees flew to New York, where he put them in a room in the American Museum of Natural History identical to the one in Paris.

The next day, at 3:15 P.M., eastern daylight time, exactly twenty-four hours after their last meal in Paris, the bees emerged from their hive and flew directly to their feeding table. Later, the scenario was reversed, with the bees trained to a New York schedule and retested on their return to Paris (chosen as the initial site because of flight schedule requirements). The results were identical.

These studies provided convincing evidence that neither unseen rays from the sun, nor any other environmental cue that varied with the earth's rotation, governed the bees' behavior. It was an internal time-keeping signal that told the bees when to stay home, and when to go out in search of food. Coming at the dawning of the jet age, the downside of this finding—biologic rhythms resist change—might have been taken as a warning. But since humans were, well, human, most paid little attention. Until they had to pay the consequences.

Like the bees, people who rapidly cross several time zones will find that their body clocks are still on home time right after they arrive. But unlike the bees, they will find many differences in their new surroundings. The sun may be shining when they feel like sleeping, or it may be dark out when they would like to begin their day. Meals will not be served when they wish to eat. Because their body clocks lag behind, or run ahead, of the new local time, they may sleep poorly and have trouble concentrating, experience slowed reflexes, irritability, gastrointestinal distress, muscle aches, and other symptoms of malaise. Some chronobiologists call this discomforting constellation of experiences *transmeridian flight desynchronosis*. The more familiar term is "jet lag."

Jet lag has caused diplomatic blunders, disrupted athletic competitions, curtailed the productivity of executives, and diminished the pleasure of travel for millions of vacationers. It may also pose physical dangers by diminishing one's coordination and driving skills for a few days, a particular hazard on unfamiliar roads. The human toll may be quite costly: nearly fourteen million Americans alone travel overseas annually.

George Bush, who vomited, fainted, and slumped to the floor while attending a state dinner in Japan in 1992, may have been suffering jet lag. He became ill on the tenth day of a 26,000-mile trip that had taken him to Australia, Singapore, and Korea before he arrived in Japan. Although the president recovered quickly, television shots of his ashen face flashed around the world, prompting new concerns about his health and ability to lead the nation, and perhaps contributing to the defeat of his bid for reelection.

In the 1950s, John Foster Dulles, then secretary of state, flew to Egypt to negotiate a treaty to construct the Aswan Dam. The United States lost the project to the Soviet Union, launching a decade of Soviet influence in Egypt. Dulles had participated in key meetings soon after his arrival; he blamed jet lag for undermining his performance, urging the nation's diplomats to avoid scheduling important meetings immediately after flying across many time zones.

United States Olympics diver Greg Louganis struck his head on a platform while diving at the 1979 Olympics trials in Moscow. It was jet lag, he said, that diminished his coordination and threw off his timing. World champion figure skater Debi Thomas fell in the first fifteen seconds of her routine at the 1988 Winter Olympics in Calgary, Alberta. The competition took place at 11 P.M., a time she normally would have been asleep. Soviet studies have shown that time shifts of as little as two to three hours can impair performance in athletes.

Jet lag has caused serious trouble in space—the ultimate time-zone disruption. On missions orbiting the earth, astronauts see a sunrise roughly every ninety minutes. During early manned space flights in the 1960s, problems with sleep and fatigue prompted changes in flight plans, although they did not necessitate the premature termination of any flights. In 1981, on the United States' first space shuttle mission, crew members slept in their seats on the flight deck, within reach of the controls in case an emergency arose. On the second night, as a test, Mission Control set off an alarm. The astronaut who responded flipped the wrong switch. The next day, he had no memory of his nighttime exchange with the ground controllers.

WHAT CAUSES JET LAG?

Pioneers traveling across the United States 150 years ago did not suffer from wagon lag. When Jules Verne and his crew sailed around the world in 1873 in a then-remarkable eighty days, no one developed ship lag. Before the invention of planes, people seldom journeyed more than six hundred miles, the approximate distance of one time zone, in a single day. Thanks to the slow pace of travel, the body's orchestra stayed in tune.

Rapid travel causes body rhythms to fall out of sync with both the environment and each other. Jet lag reflects a contest between the old time zone and the new one for their control. It is analogous to a new conductor arriving in the middle of an orchestral performance and attempting to elbow the old conductor aside. Some musicians might follow the old maestro long after the others have switched to the new one. A cacophony would ensue.

The clocks with prime importance for alertness and moods—those controlling the sleep/wake cycle and body temperature—fall back into synchrony at different rates. The clock that regulates the timing of sleep and wakefulness is fairly speedily realigned; within two or three days after a six-hour trip, most people will be able to fall asleep at their customary hour. However, they are unlikely to sleep as well as usual, because the body clock for temperature is more recalcitrant, taking about five days to recover its normal pattern. This means that body temperature will not achieve its usual low point midway through sleep, making sleep more fitful; nor will it be at its highest at the same time each waking day, causing inertia at unexpected hours.

Traveling east is harder for most people than traveling west, as it imposes a greater difference between one's usual bedtime and activities in the new time zone. When traveling east, one is required to go to bed and get up earlier than usual; this is more difficult to do, and it is more likely to curtail sleep than traveling west, where one has to stay up later and sleep later. People who travel from St. Louis to London, for example, find that when it is dinnertime by the Mississippi, it is six hours later, bedtime, by the Thames. But it will feel much too early for sleep. At bedtime in the shadow of the Arch, dawn is splashing over Westminster Cathedral.

By comparison, people who travel from Frankfurt, Germany, to Washington, D.C., find that at lunchtime by the Main it is six hours earlier, breakfast time, on the Potomac; this is less of a stretch. In general, one is sleepier than usual in the morning when traveling east, and sleepier in the evening when traveling west.

When traveling west, body clocks follow the direction of the plane. But when one travels east, it is one of the body's quirks that its rhythms do not shift uniformly in the same direction. Some slow down, while others speed up to accommodate to the new schedule. Thus, after a trip of eight time zones eastward, some rhythms will shorten by eight hours, while others will lengthen by sixteen hours. This causes greater overall dysynchrony, and it is another reason why eastward trips usually produce more severe jet lag.

German researchers Karl Klein and Hans Wegmann found that realignment of a large variety of rhythms occurred at the rate of about sixty minutes per day after flying east and ninety minutes per day after flying west. This is the source of the oft-quoted rule that adaptation takes about a day for each time zone crossed. Two or three weeks, however, may be needed to realign all rhythms completely. By then, many travelers will be back home and starting the process all over again.

If taking an around-the-world trip, most people would adapt most easily going west. Some travelers, however, especially those with depression, often prefer traveling eastward. The resulting sleep deprivation and schedule shifts are akin to those used therapeutically and may brighten their moods. Indeed, a study by British researchers P. Jahhar and Malcolm Weller of nearly one hundred travelers who arrived at London's Heathrow Airport so disturbed that they had to be taken to the nearest psychiatric hospital showed that eastward flights across several time zones sometimes triggered hypomania in those susceptible to it. By contrast, westward travel for this group sometimes brought on depression.

Jet lag is not the result of "culture shock"; people who undergo simulated time-zone crossings in the chronobiology laboratory suffer as much as any actual traveler. North–south flights that are all in the same time zone, even if the distance is equivalent to crossing several time zones, will spare the traveler many symptoms of jet lag. Sleep problems and other mood or behavior changes may develop, however, if dawn and dusk at destination occur at times that are significantly different from those at point of departure. People who visit the Antarctic in the darkness of winter may feel more depressed, and people who go to Scandinavia in the summer, when there is near-constant daylight, often stay up much later than usual.

Some discomforts that travelers experience are not caused by disruptions in body rhythms, but rather by the length and stress of the flight. Dryness of the mouth, nasal passages, skin, and eyes, for example, is caused by reduced oxygen levels and low humidity in the aircraft cabin; this symptom may be compounded by the consumption of alcohol and caffeine, both diuretics. Sitting for a long period permits fluids to drain to the lower limbs, causing the ankles and feet to swell, while cramped conditions may cause backache and stiff muscles. Additionally, carbonated drinks may produce intestinal gas and distend the abdomen; vibration, noise, air turbulence, and second-hand smoke may trigger a headache. Last-minute packing and other chores may give rise to anxiety, weariness, and irritability even before the trip begins; these reactions may not emerge until the events that prompted them have been left behind. Fortunately, however, most of these symptoms are of far shorter duration than those of jet lag, usually dissipating within a day of the trip.

SUFFERING JET LAG WITHOUT LEAVING HOME

Staying up three hours past one's normal bedtime on Friday and Saturday, and attempting to go to bed at the usual time on Sunday, will make Sunday night's sleep comparable to what one would experience after traveling from Boston to San Francisco and back. The body views the sound of the alarm at 7 A.M. Monday as a 4 A.M. reveille. The Monday morning blahs are a minidose of jet lag.

Even quite modest schedule changes, such as switching to and from daylight saving time, may induce jet lag. The spring change is harder on the body, as it is comparable to eastward travel. Timothy Monk, of the University of Pittsburgh, and Simon Folkard of the University of Sussex, England, found that daylight traffic accidents increased the week after the spring change, presumably a result of drivers suffering shorter fuses, greater fatigue, and poorer coordination.

Adjustment after daylight saving time goes into effect takes about a week, considerably longer than after a trip of just one hour, because the physical environment stays the same; that is, the body receives the same light and dark signals as before the time change. It is for this reason that farmers have long lobbied against daylight saving time. Their contention is that cows continue to get up with the sun, and they must continue to get up with the cows. But citizens of more than two dozen countries have adopted daylight saving time, opting for the pleasure of an extra hour of daylight on summer evenings.

INDIVIDUAL DIFFERENCES INFLUENCE JET LAG

The severity of jet lag varies from traveler to traveler, and in the same person, from trip to trip. About 20 percent report little trouble with jet lag, and 20 percent say they have considerable difficulty. That people at either end of the spectrum often find themselves traveling with each other is one of fate's peculiar ironies.

Early risers, or larks, have an advantage over late risers, or owls, when going east, and owls over larks, when going west. People who travel with others, particularly in groups, have less trouble with jet lag than solo travelers. The larger the group, the more likely its members will follow a regular schedule for meals and other activities, supplying numerous zeitgebers to help resynchronize their clocks. Extroverts adapt faster than introverts, presumably because they are more likely to seek out company and organized activities.

Military personnel generally adjust faster than civilians. Besides traveling in groups, the military usually spend a large part of their time outdoors, where changes in light and dark serve as the most powerful of all reentrainment cues. Additionally, they generally are young and in good health, traits linked with greater flexibility of rhythms and better sleep. They are also highly motivated to adapt, a force that distracts them from minor discomforts. People in their forties, fifties, and older usually have a harder time than younger people, chiefly because they find it difficult to adapt to the different sleep schedule required in the new time zone.

Overweight people have less trouble with jet lag than those of normal weight. This curious fact has a biologic basis: people who are overweight typically eat when it is mealtime, instead of waiting until hunger strikes. Clock-watchers usually enlist numerous external cues to speed their adaptation, whether they are aware of doing so or not. Thus, they adapt more easily to time-zone changes than those on more flexible schedules.

PILOTS GET JET LAG, TOO

Pilots are as susceptible to jet lag as their passengers. One might think that pilots could avoid jet lag simply by staying on home time during their layovers, but the differences in light and dark are too powerful for the body to ignore. Moreover, pilots must adapt at least in part to local mealtimes, and business hours, and they must contend with noise if they try to sleep during the day.

If pilots suffer less from jet lag than other people, it is only because they take preventive measures and have learned how to cope when afflicted. Studies of what pilots do and don't do when "on the road" have identified some of these strategies. Any traveler could use them to weather a trip better; they are particularly useful for business travelers, athletes, and others who must function in top form in a new time zone right after arrival.

Crew members from Pan American World Airways, British Airways, Lufthansa, and Japan Air Lines participated in a major study coordinated by the

National Aeronautics and Space Administration (NASA), conducted between 1984 and 1986.

Before and after trips across seven to nine time zones—San Francisco to and from Tokyo, London, or Frankfurt—the crew members, all male, came to the sleep laboratories at Stanford University in Stanford, California; the Royal Air Force Institute of Aviation Medicine in Farnborough, England; the DFVLR Institute for Aerospace Medicine in Cologne, Germany; and Jikei University School of Medicine in Tokyo, Japan, where their sleep and alertness were assessed. During their trips, the crews followed their normal layover routines. Like ordinary tourists, some elected to go to sleep soon after arriving at their destination even if it was still daytime, while others went sightseeing or shopping. Some stayed on home-base time; some switched to local time.

As expected, the crews who flew east slept more poorly than those flying west. The men used different strategies to obtain the sleep they needed to be alert for their next flights; some of their techniques, the researchers discovered, worked much better than others. After westward flights, afternoon naps restored alertness and improved nighttime sleep. After eastward flights, however, avoiding naps was a better ploy. The men slept more soundly if they stayed awake until the normal local bedtime.

The crew members—like most people—could not tell how sleepy they were in the daytime, said NASA psychologist Curtis Graeber, who coordinated the study. They often insisted that they were wide awake when sleep laboratory tests showed otherwise. This finding cast doubt on pilots' claims that sleepiness in the cockpit is infrequent, said Graeber, who now is at the Boeing Corporation. The study was a powerful demonstration of the value of a cookbook approach to planning one's schedule after a time-zone change: using the biologic clock to dictate daily activities assures that one will attempt to sleep at the most optimal time, and will be the least encumbered by sleepiness during waking hours.

Graeber and his colleagues advised airlines to take into account normal daily patterns of alertness and sleepiness when making flight schedules. Eastward flights from San Francisco to London ideally would depart during the late-afternoon and evening periods of high alertness, roughly 4 to 10 P.M. Thus, pilots could utilize this strong-performance period to cope with one of the most challenging parts of the flight. Such flights also would land after sunrise, when alertness is on the rise. And westward flights from San Francisco to Tokyo ideally would depart early enough in the day to arrive in Japan at night, the right time for the crew to go to sleep. (When it is noon in San Francisco, it is 5 A.M. in Tokyo; an eleven-hour flight leaving California at noon would arrive at 4 P.M. Japanese time; crew members could have dinner and go to sleep, still partly in sync with their home-base routine.) Chronobiologically sound schedules obviously would benefit passengers as well as crew. (See also Chapter 10 for more on hours of work and pilot performance.)

COPING STRATEGIES

Other than flight crews, business travelers are the most frequent flyers, logging an average of thirty-three overseas trips in a lifetime. For 40 percent of business travelers, a typical journey outside the country lasts just one to seven days. For such trips, trying to stay on home time may be the best tactic.

Some French executives follow a tightly orchestrated system to beat jet lag on their trans-Atlantic trips, according to writer Jean-Louis Servan-Schreiber. In *The Art of Time*, Servan-Schreiber relates that executives put in at least an hour's work at their Paris offices on Monday morning before heading for the airport, taking off at 11 A.M. on the supersonic Concorde. When they land in New York, it is only 8 A.M., so they can make use of a full American workday. That evening, they head back to Paris on a regular 7 P.M. flight, on which they had purchased three tourist-class seats for the same price as one seat on the Concorde. They stretch out and sleep for six hours, arriving in Paris at 7 A.M. Tuesday. This allows them to return to their offices, he asserts, "free from fatigue and jet lag." An American business executive, starting from the opposite side of the Atlantic, could not duplicate such efficiency.

HOW TO AVOID ADAPTATION

Trips of One or Two Days

To avoid adaptation, one should:

- ***Try to stay on home time, by not resetting one's watch, and following one's usual schedule as much as possible.***

- ***Stay indoors as much as possible.*** When outside, minimize the impact of daylight on body clocks by wearing dark glasses.

- ***Sleep as much as possible during the same hours one normally sleeps at home.***

On pages 168–169 is a sample schedule for a two-day business trip from the United States to England, departing—as do the majority of flights on this route—from the East Coast in the evening and returning from England in the late afternoon. The trip starts at midnight in the United States, that is, 6 A.M., British time. The schedule is on military time, a twenty-four-hour day; thus, midnight is 0, noon is 12, and 6 P.M. is 18.

Trips of Longer Duration

On a trip lasting more than a few days, readjustment is bound to occur. Daylight, mealtimes, traffic noises, the ringing of church bells . . . the myriad of time cues cannot be ignored. Accelerating adaptation would be desirable now. When preparing to fly around the world alone in 1931—a decade before the first jets streaked through the sky—aviator Wiley Post anticipated the possible adverse effects from time-zone changes and designed a strenuous training regimen for himself. In the months before the trip, he never slept at the same hours on any two days in the same week, a task he called "far more difficult than flying an airplane." Post also changed his mealtimes. These tactics, he asserted in his book *Around the World in Eight Days* (coauthored with Harold Gatty), enabled him better to withstand the rigors of his record-setting marathon flight.

Today's travelers can utilize a tamer and more effective plan based on chronobiologic principles. The following advice applies to trips of three or more time zones in either direction.

Before Departing:

- ***Start changing your sleep schedule.*** Go to bed and arise earlier if heading east, and later, if heading west.

- ***Start shifting mealtimes and daily routines such as exercise.*** While most people would find it impractical to shift their schedule completely to the time at their destination before leaving home, even small changes will be helpful. Some studies suggest that adaptation proceeds most rapidly in the first few days.

- ***Start the trip well rested.*** Sleep loss has a global impact on performance and moods, and it exacerbates other jet-lag symptoms.

- ***Consult your physician if taking any medications on a regular schedule.***

TWO-DAY BUSINESS TRIP, UNITED STATES TO ENGLAND

EASTERN STANDARD TIME (USUAL SCHEDULE)		BRITISH TIME (TRIP SCHEDULE)	
0	Bedtime; en route, sleep on plane	5	
1		6	Arrival
2		7	
3		8	Go to hotel. Nap if possible.
4		9	Avoid scheduling meetings. The next two hours are your time of lowest performance.
5		10	Stay indoors.
6		11	
7	Wake-up time	12	
8	Breakfast	13	Lunch with your British associates. Eat lightly.
9		14	Your most productive hours begin.
10		15	
11		16	
12		17	
13	Lunch	18	Have a light meal.
14		19	Nap for an hour or so if you can.
15		20	
16		21	
17		22	
18		23	Stay up late.
19	Dinner	0	
20		1	Have a light meal.
21		2	
22		3	Go to sleep, if sleepy. Try to sleep four hours or longer.
23		4	
0	Bedtime	5	
1		6	
2		7	

TWO-DAY BUSINESS TRIP,
UNITED STATES TO ENGLAND, *Continued*

3		8	If you're awake, eat lightly.
4		9	Avoid scheduling meetings. The next two hours are your time of lowest performance.
5		10	
6		11	
7	Wake-up time	12	
8	Breakfast	13	Brunch for you; lunch for your British colleagues.
9		14	Your most productive hours begin.
10		15	
11		16	
12		17	
13	Lunch	18	Have a light meal.
14		19	Nap if you can.
15		20	
16		21	
17		22	
18		23	Stay up late.
19	Dinner	0	
20		1	Have a light meal.
21		2	
22		3	Go to sleep, if sleepy. Try to sleep four hours or longer.
23		4	
0	Bedtime	5	
1		6	
2		7	
3		8	
4		9	
5		10	
6		11	Eat lightly.
7	Wake-up time	12	
8		13	Depart. On arriving home, resume your normal schedule.

Time Zone Guide

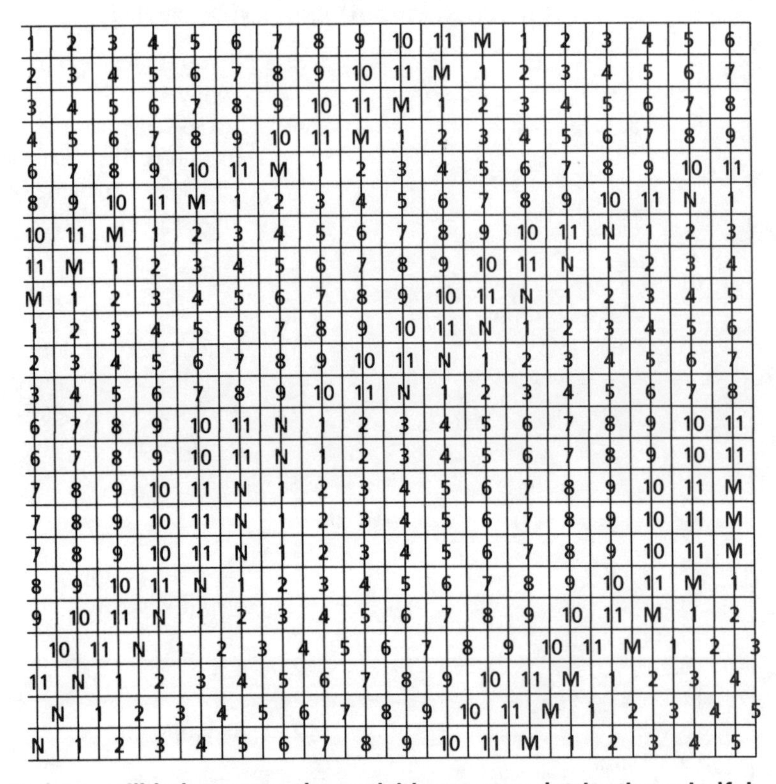

1	2	3	4	5	6	7	8	9	10	11	M	1	2	3	4	5	6
2	3	4	5	6	7	8	9	10	11	M	1	2	3	4	5	6	7
3	4	5	6	7	8	9	10	11	M	1	2	3	4	5	6	7	8
4	5	6	7	8	9	10	11	M	1	2	3	4	5	6	7	8	9
6	7	8	9	10	11	M	1	2	3	4	5	6	7	8	9	10	11
8	9	10	11	M	1	2	3	4	5	6	7	8	9	10	11	N	1
10	11	M	1	2	3	4	5	6	7	8	9	10	11	N	1	2	3
11	M	1	2	3	4	5	6	7	8	9	10	11	N	1	2	3	4
M	1	2	3	4	5	6	7	8	9	10	11	N	1	2	3	4	5
1	2	3	4	5	6	7	8	9	10	11	N	1	2	3	4	5	6
2	3	4	5	6	7	8	9	10	11	N	1	2	3	4	5	6	7
3	4	5	6	7	8	9	10	11	N	1	2	3	4	5	6	7	8
6	7	8	9	10	11	N	1	2	3	4	5	6	7	8	9	10	11
6	7	8	9	10	11	N	1	2	3	4	5	6	7	8	9	10	11
7	8	9	10	11	N	1	2	3	4	5	6	7	8	9	10	11	M
7	8	9	10	11	N	1	2	3	4	5	6	7	8	9	10	11	M
7	8	9	10	11	N	1	2	3	4	5	6	7	8	9	10	11	M
8	9	10	11	N	1	2	3	4	5	6	7	8	9	10	11	M	1
9	10	11	N	1	2	3	4	5	6	7	8	9	10	11	M	1	2
10	11	N	1	2	3	4	5	6	7	8	9	10	11	M	1	2	3
11	N	1	2	3	4	5	6	7	8	9	10	11	M	1	2	3	4
N	1	2	3	4	5	6	7	8	9	10	11	M	1	2	3	4	5
N	1	2	3	4	5	6	7	8	9	10	11	M	1	2	3	4	5

The chart above will help you to plan activities at your destination mindful of activities that ordinarily take place at the same time at home. Note that times in Iran and India vary by one-half hour from those of other countries. Some countries, including the United States, span several time zones. A nap in a new time zone will be most restorative if taken during your usual hours for sleep.

Courtesy of Charmane Eastman

Diabetics who require daily insulin, for example, must adjust doses to compensate for changes in diet, sleep, and other bodily rhythms after time-zone crossings.

- ***Ask your physician about the pros and cons of taking a sleeping pill on the plane.*** A sleeping pill may screen out many discomforts of the flight, and provide sleep not otherwise obtainable. However, it almost certainly would

7	8	9	10	11	N	1	THAILAND	
8	9	10	11	N	1	2	CHINA	
9	10	11	N	1	2	3	JAPAN	
10	11	N	1	2	3	4	AUSTRALIA (SYDNEY)	
N	1	2	3	4	5	6	NEW ZEALAND	
2	3	4	5	6	7	8	HAWAII	
4	5	6	7	8	9	10	PACIFIC	U.S.A. Standard Time
5	6	7	8	9	10	11	MOUNTAIN	
6	7	8	9	10	11	M	CENTRAL	
7	8	9	10	11	M	1	EASTERN	
8	9	10	11	M	1	2	BERMUDA	
9	10	11	M	1	2	3	BRAZIL	
M	1	2	3	4	5	6	PORTUGAL	
M	1	2	3	4	5	6	ENGLAND	
1	2	3	4	5	6	7	FRANCE	
1	2	3	4	5	6	7	DENMARK	
1	2	3	4	5	6	7	HUNGARY	
2	3	4	5	6	7	8	ISRAEL	
3	4	5	6	7	8	9	MOSCOW	
4	5	6	7	8	9		IRAN	
5	6	7	8	9	10	11	PAKISTAN	
6	7	8	9	10	11		INDIA	
6	7	8	9	10	11	N	BANGLADESH	

interfere with alertness during a midflight emergency. Further, you might arrive at your destination before its effects wear off.

Certain sleeping medications in the benzodiazepine family have been reported to cause transient loss of memory. Some travelers who took such drugs before or during a flight have related "coming to" several hours after arrival without knowing where they were or even what city they were in. In some cases, according to their traveling companions, they performed in a seemingly normal manner, even giving a lecture or participating in talks on complex topics, although they could not remember these events later.

The likelihood of amnesia after taking a sleeping pill may be increased if alcohol is consumed at the same time. According to Wesley Seidel of Stanford University, the 60 percent of Americans who drink regularly seldom resist taking a drink or two in the festive atmosphere of a trip.

- ***Factor stress into your travel plans.*** Some airports have less traffic, less distance to walk, better baggage service, and better ground transportation than others. Even if it means traveling a few minutes longer, it may be worth it to use a more amenable connecting airport.

- ***Try to break up a long flight with an intermediate overnight stopover.*** Although the total number of time zones crossed will be the same, this strategy may diminish some of the malaise associated with jet lag. Among recent presidential travelers, Ronald Reagan preferred stopovers, while Bush opted to fly straight through. When Pope John Paul II visited Colorado in 1993, American Airlines removed first-class seats and installed a bedroom on *Shepherd One*, a Boeing 767-300ER jet, for the pope's eleven-hour nonstop return flight to Rome.

- ***Start the trip on a full stomach.*** Then eat lightly until after arriving at your destination, when local mealtimes can be utilized as zeitgebers.

- ***Take along gum or candy to counteract nausea during takeoff and landing, or for turbulence, and a moisturizer to soothe dry skin.*** Although these measures do not combat jet lag per se, they can make the trip go more smoothly.

En Route:

- ***Set your watch to destination time soon after boarding the plane.*** This is a psychological step toward "doing what the Romans do."

- ***If it is nighttime at your destination, try to sleep on the plane. If it is daytime, avoid sleeping, or at least limit yourself to a two-hour nap between the hours of 2 and 4 P.M. at your destination.***

- ***Load up on water and fruit juices.*** Do not drink carbonated drinks, coffee and other caffeinated beverages, and alcohol while in the air. At high altitude, one alcoholic beverage is the equivalent of two on the ground.

- ***Take off your shoes and prop your feet on a bag or blanket to reduce swelling.*** To minimize stiffness and aid circulation, walk up and down the aisles every hour or two. Stretch, roll your shoulders, wiggle your toes, flex your ankles, clench your fists, and perform any other exercise that is possible in a restricted space. You can also utilize relaxation techniques to ease stress by leaning back, closing your eyes, and taking deep breaths.

After Arriving:

- ***Expect a certain degree of jet lag and make allowances for lowered performance.*** Some business organizations allegedly forbid their representatives to sign contracts in the twenty-four hours following a trip from the United States to Europe, and in the seventy-two hours following a trip to Asia.

- ***When traveling east, schedule important activities in the afternoon; when traveling west, schedule them in the morning.***

- ***Utilize clocks, meals, social activities, and exercise as zeitgebers.*** Eat on local time, even if you are not particularly hungry. Eat lightly for the first couple of days to minimize gastrointestinal distress. Exercise on local time, but take it easy to avoid injury.

- ***Spend as many daylight hours outdoors as possible.*** Travelers who spend time outdoors adapt twice as fast as those who do not. At the local time for sleep, rest in a dark room, even if you are not sleepy.

- ***Try to get at least four hours of sleep during your usual time period at home.*** This maintains some stability while realignment of your rhythms is in progress. Eleven P.M. to 3 A.M. in Washington, D.C., is 5 A.M. to 9 A.M. in Amsterdam, for example, an appropriate sleep time in both time zones. If you are feeling sleep-deprived, take a short nap in the afternoon.

At the beginning of an eastward trip, good nights often alternate with bad nights. Most people stay up late to pack and finish last-minute chores. On the plane they seldom sleep as long, or as soundly, as usual. The resulting sleep debt helps them to sleep soundly the first night at their destination, even though their body clocks know it is the wrong time. The second night, they are not overwhelmingly sleepy, and sleep restlessly. On the third night, sleepiness again surges to the fore, the readjustment of body clocks has begun, and the quality of sleep is better. By the fourth night, sleep is nearly normal.

Those who travel for business rather than for pleasure may find sleeping pills helpful for a few nights. It may seem paradoxical, but the best time to take a sleeping pill is the night after you have slept well, since that is when sleep is most likely to be disturbed. Rapidly eliminated medications are generally advised for fighting jet lag; they induce sleep, even though it is the wrong time on the biologic clock, but they usually do not stay in the body long enough to cause any disorientation the next day. It is prudent to try any medication at home first before taking it on a trip.

- ***Drink caffeinated beverages only when you are feeling sleepiest.*** This will maximize caffeine's alerting effect.

- ***Do not drink more than one glass of beer or wine near bedtime.*** As the alcohol wears off, sleep will be disrupted.

On Returning Home:

- ***Expect a certain degree of jet lag; make allowances for lowered performance.*** Some people feel they adapt faster when they return home, regardless of whether home is east or west of where they have been. However, whether your flight is outbound or homebound, in the daytime or at night, does not alter your body's ability to adapt. A return to familiar surroundings and regular routines may just make jet-lag symptoms less noticeable.

SHIFTING BODY RHYTHMS

The discovery of medications with a chronobiotic effect—that is, the ability to move daily rhythms around—would have obvious applications in the treatment of jet lag. While there are no drugs available whose only effect is chronobiotic, some sleeping pills may serve this purpose, recent studies suggest. Bright light and a special diet also may help counter jet lag.

Effects of Sleeping Pills

A person traveling westward wants to sleep earlier; one who is traveling eastward wants to sleep later. A sleeping pill's efficacy in treating jet lag may depend not only on its ability to induce sleep, but also on its ability to induce sleep at the desired time, and perhaps to move other circadian rhythms earlier, or later, as well.

Animal studies suggest that sleeping pills may help alter habitual daily behavior. In one study, a commonly prescribed medication, triazolam, helped promote changes in wheel-running by golden hamsters, behavior that typically varies by fewer than fifteen minutes from day to day. Fred Turek and Susan Losee-Olson of Northwestern University first plotted the habitual patterns of each of a group of hamsters, then put them in a room where the lights remained on or off constantly for at least two weeks to eliminate either light or dark as a zeitgeber. The researchers then injected the hamsters with either the triazolam or an inert solution at different intervals before and after the animals' usual wheel-running time. A single injection of the drug caused the hamsters to become less active at first but markedly more active later. The next day the hamsters that received the drug before their usual wheel-running time ran earlier; those that received it afterward ran later. The inactive solution had no effect on them. This study suggests that the withdrawal of the triazolam, rather than its consumption, brought about the behavioral changes. It also suggests that activity per se may be sufficient to shift rhythms—an appealing idea since exercise is easily rescheduled. In a later study, Turek and Olivier van Reeth of the Free University of Brussels found that a single injection of triazolam made hamsters adapt twice as fast as they would otherwise to changes in routine that mimicked an eight-hour trip eastward.

A related study, conducted by Peretz Lavie of the Technion-Israel Institute of Technology in Haifa, showed that a similar drug, midazolam, altered body temperature, perhaps the most stable—and hardest to budge—of all daily rhythms. In volunteers who traveled west after taking the drug, body temperature shifted earlier. In those traveling east, temperature shifted later.

The hormone melatonin may also be a jet lag remedy. As melatonin is secreted principally in the dark, researchers theorized that if one took it in tablet form at the appropriate time one might fool the body into thinking that darkness had come earlier, or later, thus causing the melatonin secretion, as well as other rhythms, to shift in the same direction. In one study, seventeen people flew from London to San Francisco, and then spent two weeks in the United States, taking either melatonin or a placebo for three days before returning home and for four days afterward. All who took the melatonin escaped jet lag, but most of those who took the placebo suffered from it, according to Josephine Arendt and her colleagues at the University of Surrey in Guildford, England.

Bright Light May Speed Adaptation

A properly timed dose of bright light may be able to twirl the hands of the biologic clock faster than simply shifting sleep and other daily activities would.

In one test of this theory, Mike Long, a journalist returning to the United States after a month in the Far East, donned welder's goggles to shield his eyes at the first suggestion of dawn on the flight from Tokyo to Boston. Once in Boston, Long headed for the chronobiology laboratory at the Brigham and Women's Hospital, where researcher Charles Czeisler kept him isolated from all time cues, exposing him to bright artificial lights for three days in a row when Long's temperature was lowest. The result: Long's temperature rhythm shifted from Tokyo time to Boston time, a difference of eleven hours, in three days. This is roughly four times faster than is usual. "My jet lag vanished in the lights," Long reported in the *National Geographic*.

This experiment was part of an effort by Czeisler and his colleagues, as well as chronobiologists elsewhere, to map the human phase response curve, the biologic rhythm of responsiveness of body clocks to light exposure. Phase response curves, or PRCs, have already been determined for many animals, and they are all pretty much alike, leading to the conclusion that the human pattern must follow a similar path.

Exposure to a pulse of light at different times of day can trick animals into responding as if the sun had risen or set at a different time. The time the light exposure is received is crucial. Light in the morning advances such body rhythms as physical activity, hormone secretion, urine output, and body temperature rise, while evening light makes the same events occur later. Light in the middle of the day seems to have no effect. The transition point on the PRC occurs about 5 A.M., or the time of lowest body temperature.

Since sunrise and sunset actually do change relative to sleep when one crosses time zones, it seems logical that exposure to light—or to darkness—at the right time might help restore the normal relationship of these events. Shifting human rhythms requires much brighter light, and much longer exposure, than is the case for laboratory animals who never see daylight.

Richard Kronauer, a mathematician working with Czeisler, has produced a theoretical model of how light alters human rhythms. He and Czeisler used this model in studies simulating jet lag in the laboratory, finding that light brought about the most substantial shifts in rhythms when body temperature was at its daily low.

They found that exposure to bright light when temperature was lowest on three successive days could accelerate a rhythm shift in their subjects. On the first two days exposure to light jolted the biologic clock and blunted the range of normal daily highs and lows. After the subjects "traveled" to a new time zone by changing their times of sleeping and waking, the third exposure to light served to restart the clock at the desired time. To use the researchers' analogy, the first two exposures will send a person to the North Pole, while the third will take him or her to the desired time zone. On a trip, then, one should spend at least sixty minutes and preferably longer outdoors on the first two days, particularly around the middle of one's normal sleep time at home when body temperature is at its lowest. Arising at local time on the third day and going outside soon afterward should lock one into the new time zone.

Much about the human PRC still remains unknown. If light is as potent a tool as laboratory studies suggest, there is a danger that a dose given at the wrong time could send someone chronobiologically in the wrong direction. It would be as if a person traveling from the Midwest to Italy ended up in Hawaii instead.

Scientists still need to ascertain not only when to give light but how intense it must be, how long a dose is necessary, and how many days of exposure are required as well as to determine the body's receptivity to light at different times of day. Also, they must show that light exposure that can shift rhythms under laboratory conditions can do so in the real world, where it must compete with natural light and dark cues. These questions are similar to those involved in the use of light to treat seasonal affective disorder and various sleep problems, and to help shift workers cope better with schedule changes.

Although the use of light to prevent or cure jet lag is still in its embryonic stage, it is possible to imagine that in the future, one will start the process at home before a trip by going outdoors at the right time of day or by using bright artificial lights. On the plane, flight attendants may hand out computerized light visors into which travelers will punch their own schedules. At the appropriate time, each traveler will be told, "Turn on your lights in preparation for landing."

Does the Jet-Lag Diet Work?

The theory behind the much publicized jet-lag diet is that one can reset one's biologic clock by eating certain types of foods at specific times. The "world traveler's diet," developed by biologist Charles Ehret at the Argonne, Illinois, National Laboratory, grew out of animal research suggesting that protein foods promoted wakefulness, while carbohydrates promoted sleep, findings borne out, to a modest degree, by studies in small numbers of human subjects. Ehret also proposed that timed doses of coffee and tea, properly viewed as psychoactive drugs, could help reset rhythms.

Ehret has described his diet in a book, *Overcoming Jet Lag*, coauthored with Lynne Waller Scanlon. The regimen takes a good amount of diligence to follow scrupulously. One must start it four days before takeoff, counting backward from breakfast time at one's destination on the day of arrival:

Day 1: "Feast" with three full meals. Breakfast and lunch should be high in protein, with such foods as eggs, chicken, and fish. Dinner should be high in carbohydrates, with such foods as pasta and potatoes. No coffee except between 3 and 5 P.M.

Day 2: "Fast" with three small low-calorie, low-carbohydrate meals, such as salads, light soups, fruits, and juices. Again, no coffee except between 3 and 5 P.M.

Day 3: Repeat Day 1.

Day 4 (day of departure): Repeat Day 2.

Caffeinated beverages should be consumed before noon when traveling west; after 6 P.M. when traveling east. On the plane, pass up the dinner and try to sleep. Have a high-protein breakfast.

But does this regimen really make any difference in the speed of one's adaptation? The diet attracted attention in the early 1980s when it was used in studies of jet-lag countermeasures taken by military troops being deployed overseas. The diet was, however, only one of several strategies the troops followed. They also adhered to a regular sleep and activity schedule that may have been as important as, perhaps more important than, the diet itself in helping to speed their adaptation.

To date, there has been only one laboratory study of the diet alone in human subjects. Cornell University researchers found that it conferred no special advantage. Margaret Moline and her colleagues at Cornell used an established jet-lag simulation routine for their study. Fifteen healthy young adults, who were isolated from time cues in the chronobiology laboratory, stayed on their usual schedules for a week. Their schedules were then changed to mimic a flight to Europe, and they remained in the laboratory a week longer. Seven of the participants followed the jet-lag diet, while the rest consumed their usual diet. After the time shift, both groups slept equally badly, and the rate of adjustment of body temperature, mood, and performance was also identical.

Thus, there is no proof yet that any diet hastens resynchronization. But the psychological benefits of "doing something" probably should not be trivialized, and the possibility that certain foods can help combat jet lag—certainly a more innocuous strategy than using drugs—remains promising.

SUMMING UP

Once thought to be all in the mind, jet lag is now recognized as a persuasive demonstration of the profound *dis-ease* a person may suffer when the body clock falters. Thanks to chronobiology studies, many pieces of the jet-lag puzzle have fallen in place, and the solution to this vexing problem has started to emerge. The answers are not only a boon to travelers for whom jet lag is a temporary inconvenience but also for the large and rapidly growing proportion of the world's population who work around the clock, for whom jet lag has become a chronic way of life.

CHAPTER 10

WORK AROUND THE CLOCK

SUPPOSE ONE WERE TO FLY from San Francisco to London, spend a week there, fly to Tokyo for another week, then head back to San Francisco, and stay there just one week before repeating the journey. One would be chronically jet-lagged. Workers who change from a day shift to an evening shift to a night shift make an equivalent journey through time. Their jet lag is typically worse than that of most travelers.

For travelers, the major zeitgebers—changes in light and dark, sleep times, mealtimes, and other daily activities—encourage adjustment. For shift workers, the very same cues discourage it. People who leave work at 7 A.M., for instance, find big yellow schoolbuses on the road, restaurants serving pancakes, and bars and movie theaters closed. On radio and television, they hear fast-talking morning newscasters, and traffic reports. The disharmony between their inner world, where the day is winding down, and the outside world, where the day is just beginning, markedly delays their adaptation.

In people who fly across eight time zones, most rhythms will fall back in line in approximately eight days, but shift workers who rotate eight hours may take three weeks, or even longer, to regain their synchrony. Yet shift workers typically remain on a shift for just one week. Moreover, on their days off, even when working at night, most return to the habits of the conventional daytime world, staying awake during the day and sleeping at night. They thus live with their body clocks perpetually out of sync.

The repercussions are appalling, for both individuals and society: shift workers have more health problems than their counterparts on straight day shifts; they have more injuries, more industrial accidents, and more motor vehicle accidents. They also have a higher rate of job turnover, poorer morale, and more family problems. Martin Moore-Ede of Circadian Technologies, Inc., a shift work research and consulting firm in Cambridge, Massachusetts, has estimated that human maladaptation to around-the-clock operations costs the United States more than $70 billion a year in lost productivity, lost wages, medical expenses, property damage, and insurance costs.

Some twenty million Americans may suffer from chronic occupational jet lag, for one in three working men and one in four working women regularly works hours other than nine to five. A national household survey conducted in May 1985 for the Bureau of Labor Statistics found that Americans worked an astonishing 576 different schedules. One sixth of full-time workers, and one half of part-timers, usually have jobs that are outside the typical daylight working hours.

For example, Federal Express and other overnight delivery services conduct the bulk of their business at night. The Philadelphia Stock Exchange opens at 4:30 A.M. eastern standard time to give European investors access to the exchange's currency markets. The New York Stock Exchange is exploring the possibility of around-the-clock operation.

Some jobs, such as those involving food and public transportation, often require split shifts because of the uneven demand for such services. Other jobs involve a mix of days, afternoons, and nights in no particular order. Police assigned to the vice squad, for example, may not work the same shift for even two days in a row to avoid alerting those who are being investigated.

Pity the astronauts. Until recently, while in space, they lived on Mission Elapsed Time, which started with liftoff, whenever that was, no matter what time it was on their body clocks when their trip began. Ordinarily, they would have awakened around 2 A.M. to suit-up for a departure at 7 A.M.—unless, as often happened, delays kept them waiting on the launchpad. Once aloft, on dual-shift missions, the astronauts on the night shift were supposed to go to sleep immediately. Seventy percent of the crew members ended up using sleeping pills. Even so, almost all returned home sleep-deprived. One astronaut told Harvard researcher Charles Czeisler: "We spent years of arduous training for the privilege and pleasure of looking at the earth from space. When you're on duty, you're too busy to look out the window. When you finally get the time to do so, you don't want to waste it by going to sleep."

Growing knowledge about biologic clocks has brought about improvements in the astronauts' schedules, with more attention now paid to coordinating flight plans with their body rhythms. During the eleven-day mission to repair the flawed Hubble Space Telescope in 1993, the seven-member space shuttle crew worked only at night—a time dictated by Hubble's own orbit. Although the astronauts' rhythms had been preset to this schedule, with daily doses of bright lights and shifts in their sleep times and mealtimes, the nightwork in the sky induced jet lag in many who stayed home: the ground controllers and staff at the Johnson Space Center in Houston, Texas, had to work the same hours as the astronauts, and millions of television viewers sacrificed sleep to watch live broadcasts of the five lengthy spacewalks.

Those who work in human chronobiology laboratories—whose mission is to keep their subjects from guessing the time—put themselves on the worst schedules imaginable, working for varying amounts of time, and at different times of day, from one day to the next. The nation's two million nurses have considerable latitude in designing their own schedules, but they typically base their decisions on social criteria, such as blocks of time off, without taking their body rhythms into account. Indeed, it is ironic that health care personnel in general often have the worst work schedules of any workers in this country.

THE THREAT TO PUBLIC HEALTH AND SAFETY

Rembrandt's famous painting "The Night Watch" shows a military company deep in shadow, wide-eyed and attentive. Rembrandt's portrayal, while dramatic, does not jibe with reality. In the middle of the night, soldiers not in battle are more likely to be yawning and nodding. From the Trojan Horse to Pearl Harbor, attackers have often found it advantageous to strike during the sleepiest hours of the day, catching their enemies off guard. In a world where wars are increasingly fought around the clock—dramatically emphasized in American homes by nonstop television reports during the 1992 Persian Gulf war—soldiers must face two enemies, one of which is a decline in alertness at night.

In horse-and-buggy days, one could fall asleep with the reins in one's hand, confident that Old Nelly would amble home safely. However, today workers do less with their hands and more with their heads. They are less apt to operate machines than to monitor them. And the consequences of their mistakes are much higher.

Torbjörn Åkerstedt of the Karolinska Institute in Stockholm wired train drivers with portable sleep monitors. He found that the men could "drive" the train, apparently semiautomatically, for some minutes while falling asleep. They either closed their eyes or stared straight ahead. In some cases, they had obvious performance lapses, such as driving against a red light, without being aware of them. They also did not realize that they were falling asleep, although they had rated themselves as sleepy shortly before such events.

At 7:54 A.M., on January 14, 1988, two freight trains plowed into each other near Thompsontown, Pennsylvania. The engineers and brakemen on both trains were killed. Damage to the trains exceeded $6 million. None of the crew members on the westbound train had slept more than two hours in the preceding twenty-two to twenty-four hours. The probable cause of the accident, according to the National Transportation Safety Board (NTSB), was the men's inability to stay awake and alert.

"Human error causes two thirds of all workplace accidents," David Dinges of the University of Pennsylvania testified at hearings of the National Commission on Sleep Disorders Research in 1990. "A key cause of human error," Dinges said, "is inadequate sleep; it is at least as important as drugs, alcohol, and equipment failure." Inadequate sleep is not, as is commonly held, simply a matter of volition—lifestyle or poor judgment. It is a predictable consequence of living and working on schedules that defy the biologic clock.

In a recessionary economy, neither manufacturers nor retailers want to keep merchandise in warehouses. This economic climate has fostered the recent widespread practice of "just in time" delivery. As a result, truckers are under tremendous pressure to cover as much ground as possible in as short a time as possible. Trucks represent only 10 percent of all vehicles on the road, but they account for nearly one third of all motor accidents. In a one-year survey of every accident in eight states in which a driver of a heavy truck was killed—about one quarter of all such accidents nationwide—the NTSB found that fatigue was the most frequent probable cause. Fatigue figured in 31 percent of the fatalities, ahead of alcohol and other drug use, which was cited as the probable cause of 29 percent. In many accidents, both factors played a key role.

"Biologic rhythms had been taken into little account before Three Mile Island," Hugh Thompson of the United States Nuclear Regulatory Commission told a congressional panel investigating shift work in 1983. Few warnings could be more ominous than the near meltdown of a nuclear power plant. Yet inattention to body rhythms has continued to jinx the nuclear energy industry. In 1987, the Nuclear Regulatory Commission closed the Peach Bottom plant in Delta, Pennsylvania, after finding control-room operators, particularly those on the night shift, sleeping on the job.

Workers in low technology fields are equally prone to errors of oversight when fatigued. In 1987, a boatswain on a British ferry slept through the call to close the loading doors. Nearly two hundred people drowned when the ship capsized off the Belgian coast.

Machines work as well at 4 A.M. as at 4 P.M., but people do not. "No responsible manager would consider operating a piece of machinery outside its design specifications," Martin Moore-Ede has asserted. "That would lead to excessive wear, frequent breakdowns, and early replacement," he said. "Yet managers and workers alike have accepted as inevitable the physiological costs of shift work schedules that exceed the design characteristics of the human circadian system."

POOR SCHEDULES AND PILOT FATIGUE

It was 1 A.M. local time on September 1, 1983, when Korean Air Lines Flight 007 started to stray off Red Route 20, a navigational highway between Anchorage and Seoul. By 3:30 A.M., the Boeing 747 was more than three hundred miles off course, in Soviet airspace. Viewing the plane as a military intruder, the Soviets shot it down, killing all 269 persons aboard. The plane's flight-data recorder and cockpit voice recorder, the "black boxes" (actually painted orange) whose recovery had been kept secret until 1993 when Russian President Boris Yeltsin ordered them handed over to the International Civil Aviation Organization, a United Nations agency, reveal a listless, yawning crew. Even after an inquiry from another Korean airliner supposedly only thirty-two miles away about strong winds that 007 had not encountered, 007's three-man crew did not check the route. A navigational error clearly was made—a human error—but none of the sleepy crew caught it.

International flight schedules require United States–based pilots to work through their normal sleep period about 50 percent of the time. On domestic routes, that happens about 10 percent of the time, according to Delta Air Lines Captain Richard Stone, speaking for the Air Line Pilots Association, a pilots' union.

Most pilots work variable hours. With a good schedule on domestic flights, Stone said, a pilot consistently works the same shift during a typical three-day sequence. A bad schedule—bad because it generates more fatigue—involves a rapid shift of duty cycles from day to night to day. After a night flight, the crew often has to cut short the following night's sleep to report for duty in the early-morning hours on the third day.

Fatigue, while hard to quantify, is an "accident enabler" in the jargon of human-factors researchers. Charles Lindbergh nearly ended up in the Atlantic when he dozed off briefly during his historic New York-to-Paris flight in 1927. Pilots themselves blame fatigue for more than 20 percent of the close calls and other incidents they report to a voluntary, confidential call-in program started by NASA in 1985 to flag risky crew behavior, and to alert pilots to potential problems, according to NASA psychologist Mark Rosekind. The program receives more than three thousand calls a month.

Here is a typical—and to a hapless passenger, frightening—report from the captain of a wide-body commercial jet airliner: "Landed at [a large Midwest] airport without landing clearance, discovered when switching radios to ground control. . . . The primary factor was crew fatigue caused by late night departure [0230 local time], inability to rest prior to departure [both pilots attempted afternoon naps without success], long duty period [11.5 hours' duty time with en-route stop at JFK]. . . . Both pilots were having trouble throughout the let-down and approach phases of this last leg of the trip remembering altitudes cleared to, frequencies, and even runway cleared to."

Investigators trying to sort out causes for an airplane accident seldom point a finger at fatigue alone. "It is rarely a single thing that goes wrong. It is typically a chain of events," said psychologist John Lauber, former director of the NTSB.

On flight simulators, pilots make more errors, and bigger ones, in the small hours of the morning. Calls to NASA's reporting system show that problems peak between midnight and 6 A.M. During these hours, pilots, like other people, drift more easily into "microsleeps," bursts of sleep that last only a few seconds but disrupt attention. On a flight from Buenos Aires to Madrid, crew members wore portable brain-wave monitors. All five crew members lapsed into sleep between 4 and 6 A.M., while the plane was cruising over the Atlantic on autopilot.

Crew coordination, research has suggested, can cut the impact of fatigue. After crew members spend several days flying together, they work more collaboratively. Despite their fatigue, they perform better on a flight simulator than do rested, preduty crews. Their collaboration is particularly apparent in such important tasks as planning the landing approach or responding to an emergency.

"Most accidents result from crews' not doing what they're supposed to do at the appropriate time," said Federal Aviation Administration (FAA) psychologist Clayton Foushee, who directed the flight simulator studies. Because of small-group dynamics, subordinates sometimes hesitate to question their superiors. Some captains have ineffective managerial styles, Foushee said. They may be too overbearing or not assertive enough in taking charge. FAA regulations now mandate that all major carriers provide management training, including simulator exercises involving complicated problems requiring cooperation.

Today's pilots punch fewer buttons and watch fewer dials than their predecessors did. In place of a steering wheel, they now have a sidestick controller, a single lever used much like a joystick in a video game. Television screens tell them what their planes are doing. The highly computerized cockpits, known as "fly by wire" or "fly by light" systems, have prompted concerns about the impact of fatigue, and time of day, on vigilance.

"There's a tendency to believe that automation will do away with human error," Lauber said. "If the plane is automated from takeoff to touchdown, people fall into a trap. They assume things will work, and they may not be prepared when the usually reliable system doesn't work."

On February 19, 1985, when a China Air Lines jumbo jet lost power in one engine near San Francisco, the plane was operating on autopilot. Focusing on the engine problem—not considered an emergency since the plane's other three engines would ordinarily suffice for flight—the pilot did not notice that the plane was rolling to the right. The aircraft tipped over and plummeted six miles in three minutes, mostly upside down, before he was able to stabilize it. While he managed to land safely, 2 of the 274 people on board suffered serious injuries and the plane had major structural damage.

"Had the captain placed himself in a 'hands on' relationship with the airplane by disconnecting the autopilot at the onset of the engine problem," the NTSB report said, "he probably would have been more alert to the increasing asymmetrical forces being exerted on the airplane." The probable cause of the accident, the NTSB concluded, "was the captain's preoccupation with an in-flight malfunction and his failure to monitor properly the airplane's flight instruments which resulted in his losing control of the airplane." According to John Lauber, there is a "high probability" that both fatigue and time of day undermined the pilot's performance. The incident occurred ten hours after the flight began, and four to five hours after the pilot's usual bedtime, about 2:14 A.M. Taiwan local time.

"One problem with automation is that we've created a task that is deadly dull, and I say 'deadly' advisedly," Lauber added. "We need to design cockpit systems to take advantage of the benefits they can provide, such as fuel conservation, but we need to build in active involvement of crew members in the task at hand. Ideally, we'd use computers to monitor how people perform, not the other way around.

"Computers," Lauber continued, "do not respond to unforeseen circumstances, while people are good at creative real-time problem solving." Quick action by cockpit crews saved hundreds of lives when a cargo door tore off a 747 jumbo jet near Honolulu on February 24, 1989, and again, when a DC-10 lost hydraulic power near Sioux City, Iowa, on July 19, 1989. In another incident on December 15, 1989, all four engines of a 747 jumbo jet shut down temporarily when the plane flew through a cloud of ash from the erupting Redoubt Volcano

Four Days with an International Flight Crew

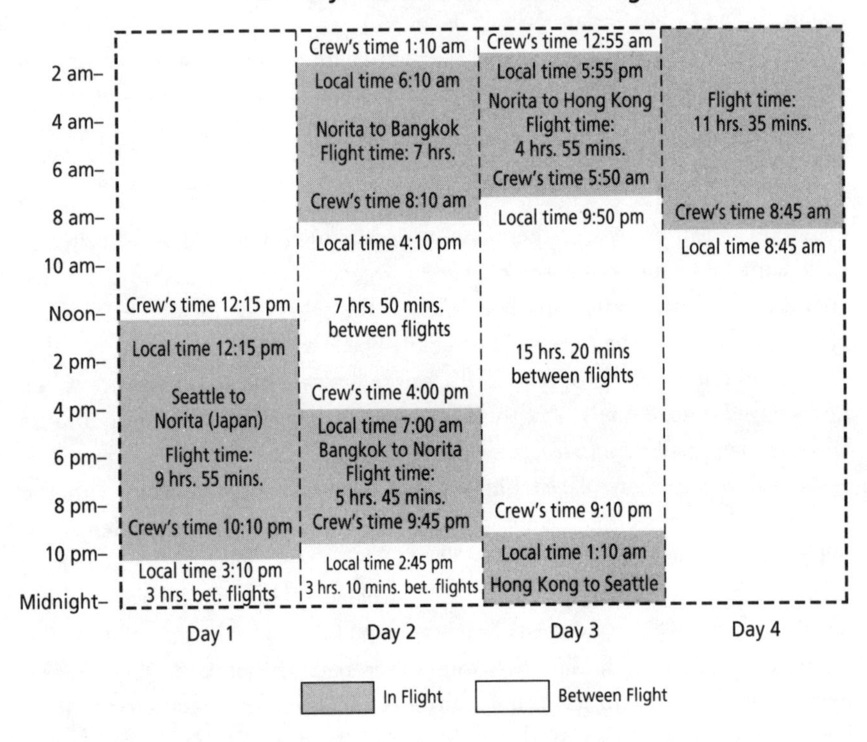

For a long trip such as this one, at least three pilots and a flight engineer would be on board. One of the pilots would also be qualified as a flight engineer. Two pilots and a flight engineer would be on the flight deck at all times, giving the other crew member(s) a chance to rest.

Data from Bill Edmunds, Air Line Pilots Association. In *Biological Rhythms: Implications for the Worker*. United States Congress, Office of Technology Assessment. OTA-BA-463. Washington, D.C.: Government Printing Office, September 1991, p. 75.

in Alaska. Although the plane plunged thirteen thousand feet in eight minutes, the crew was able to restart two of the engines and land safely in Anchorage. No one was hurt.

"A computer can fly an airplane better than the world's best pilot," said Foushee. "The driving force has been to make machines as efficient as possible, but this may not be the most efficient way to keep humans involved."

THE ORIGINS OF SHIFT WORK

Centuries ago, guards at military camps and city gates, shepherds in the field, and sailors at sea took turns at night watch. Doctors and nurses ministered to the sick around the clock. But their numbers were small.

As civilization advanced, night work increased. The streets of Rome in the first century A.D. grew so congested that Julius Caesar restricted wheeled traffic to the nighttime. "In what rented apartment is sleep possible?" the poet Juvenal grumbled a few years later. "The passage of vehicles in the turns of the street, the curses of the muledrivers who are not moving," he wrote, "deprive it even from Emperor Claudius himself."

Once the Industrial Revolution got under way in the late 1700s, many refineries and foundries began to keep their fires going around the clock. But the vast majority of workers—men, women and even young children—still worked in the daytime and slept when it was dark. William Murdock's invention of coal-gas illumination made the modern concept of shift work possible. Gas lights, first used in 1803 to illuminate the Soho Steam Works of England, rapidly replaced candles and oil lamps. Kerosene lamps appeared in the middle of the nineteenth century. Soon after night work started in the mills of New England, horse-drawn night lunch wagons began to appear on the streets. In 1882, Thomas Edison invented the electric light, sparking a vast increase in both new services and, thus, in the numbers of people to provide them.

Early in this century, workers at iron foundries and steel mills put in twelve-hour days, seven days a week, alternating between days and nights every two weeks. Once every twenty-eight days, some workers had to stay on the job for twenty-four hours straight, while others got that day off, their only day off in that four-week period.

So onerous were these schedules that the rise of labor unions had to have been entirely predictable. Religious and civic leaders also voiced concerns about the demands placed on workers, and the length of the workday slowly shrank. In 1916, the Adamson Act established the now-familiar eight-hour day in the United States; the Act did not forbid longer workdays, but provided that eight hours was to be the basis for calculating overtime. The passage of the Fair Labor Standards Act in 1938 mandated the still-standard forty-hour week, with premium pay, time and a half, for overtime. This provided employers with a further economic incentive to limit work schedules.

The most common shift-work schedule in the United States still runs for twenty-eight days: seven on the day shift, followed by two days off, seven on the night shift with two days off, and then seven on the evening shift with three days off. Such schedules require four crews to fill all three eight-hour shifts, yet give some workers a break.

The eight-hour day is now somewhat in flux. Twelve-hour shifts are making a comeback. Some hospitals provide a full week's pay for nurses who work two twelve-hour weekend shifts. The premium pay suggests that most nurses consider that these hours are undesirable; they rob nurses of their weekends, and thus interfere with their social life. On the other hand, such schedules are attractive to two-income families with young children at home, because they maximize the amount of time that at least one parent can provide child care.

Twelve-hour shifts are growing in popularity for workers in nuclear power plants, and in the chemical and petrochemical industries. Workers typically cycle between three days of work with four days off, and four days of work with three days off, with hours running from midnight to noon or noon to midnight. Many like such schedules, because they provide fourteen days off out of every twenty-eight, usually including every other weekend. Employers prefer twelve-hour shifts because they mean fewer employees, fewer schedules, and fewer records to keep.

Firefighters ordinarily work twenty-four-hour shifts. On busy days, they may not stop long enough for sleep. United States Navy submarine crews typically follow a schedule of six hours on duty and twelve hours off, or an eighteen-hour day/night cycle. While the intent is to keep the crew from performing arduous work for more than six hours in a row, this schedule is like taking a six-hour flight every day, and it is no surprise that one third to one half of the crew members do not sign up for another voyage. Unlike dayworkers, shift workers work on Thanksgiving and the Fourth of July. They also work extra shifts to cover for absent co-workers. In the chemical, food, and other industries that must run continuously to keep products from deteriorating, it takes four forty-two-hour weeks to staff the 168 hours in a week. As a result, the average rotating-shift worker puts in four hundred hours more a year than the average dayworker, or the equivalent of ten additional forty-hour weeks. Shift workers earn about 7 to 10 percent more per hour than straight dayworkers—a high incentive to stay on the job despite the risk to their health.

Many shift-work schedules followed today are an anachronism, experts in the field say. "Work schedules are inherited from generation to generation, copied from nearby plants, or negotiated," according to shift work consultant Richard Coleman of Ross, California. Schedules may not come close to matching the organization's workload. Some police departments have equal numbers of officers on all three shifts, for example, although most crimes occur during the evening shift. Many utilities do the same, even though the highest usage of their services occurs during the day. "Most companies start with a schedule that was not designed for their plant, and then develop a complicated series of rules to make that schedule work for them," Coleman said.

HEALTH PROBLEMS LONG IGNORED

Recognition of shift-work–related health problems has begun to surface only recently. It is now apparent that some problems develop slowly and only after people have been many years on the job. Other factors also have conspired to keep the issue submerged. Workers who might be expected to have the most severe health problems—those who find their schedules intolerable—often switch to other jobs, thus removing themselves from study. The extent of health problems in older workers has been underestimated, because seniority has allowed many to opt out of shift work or brought promotions to less grueling schedules.

Physicians may have missed seeing the pattern of health problems arising from shift work, because shift workers visit their doctors less often than dayworkers. It may be true that they have less free time when doctors' offices are open, but there is a more baleful reason: many are fatalistic about the problems that their schedules cause. They assume that nothing can be done. It is possible that workers who stick with shift work, those who can tolerate frequently changing schedules, are also the ones who are least likely to have health problems. That may be a Pollyannaish view, however. Some problems may have gone undetected because doctors have not been educated to make the connection between these problems and work schedules. In the professional literature of industrial medicine, exposure to toxic chemicals has stimulated thousands of articles; exposure to toxic schedules, virtually none.

The influence of social class on public concern must not be ignored. There was less call for schedule improvements when virtually all who toiled at odd hours were blue-collar workers. Indeed, when France first considered placing limits on night work at the turn of the twentieth century, some of the objections raised were gastronomic. *Quelle idée!* If the bakers did not work at night, the morning bread would not be fresh.

In the United States, the civil rights movement of the 1950s boosted activism in the workplace. The increased importance of global communications brought a new breed of shift workers, in health, managerial, or other professional occupations, who are more vocal about identifying problems and demanding solutions. In this country today, about 10 percent of shift workers are in professional occupations, and this proportion is growing rapidly. Between 1978 and 1985, night work by technical and office personnel increased 36 percent. The number of clerical workers working at night expanded three times faster than the number of all other night workers.

HEALTH HAZARDS FOR INDIVIDUAL WORKERS

As society moves toward a twenty-four-hour world, "shift workers are the advance team," Gary Richardson of Harvard University has suggested, adding, "They will be the first to show the problems of chronic disruption of body clocks." These problems may be quite substantial, undermining the quality of life, and perhaps even life itself. Among them:

- ***Sleep is shorter and less restful.*** About two thirds of shift workers complain of disrupted sleep; one third of dayworkers have such complaints. Shift workers say they cannot sleep as long as they want, and average at least an hour less sleep every day than their dayworking peers. Experience of many years does not bring relief, and even permanent night workers seldom sleep as long, or as well, during the day as at night.

  Shift workers often blame noise for their disturbed and shortened daytime sleep, but it is the natural alerting rhythm of the human circadian system that is to blame. When body temperature is rising or high, it is harder to fall asleep, and sleep is shorter. Studies in soundproof rooms show that daytime sleep is always more precarious than nighttime sleep. Noise unquestionably makes it harder to sleep. The average household is a terrible place to sleep during the day: children play, the vacuum cleaner roars, the dishwasher hums, the doorbell and the phone ring, and the noise of traffic, planes, and ordinary conversations are hard to screen out.

  "The sleep of the shift worker is not as protected by society's taboos as that of the dayworker," Timothy Monk has observed. "No one would think of phoning a dayworker at 2 A.M.," he said. "Few would have qualms about phoning a night worker at 2 P.M." Moreover, dayworkers do not have to sacrifice sleep to see the dentist, the mortgage-loan officer, or their child's teacher, while night workers often do.

  When working in the evening or at night, shift workers seldom follow the dayworker's pattern of work, leisure, sleep. Most tumble into bed right after work. For the typical dayworker, this pattern would be analogous to leaving the work place at 5 P.M. and going to sleep at 6 or 7 P.M. Before the first night shift in a sequence, workers may stay awake for twenty-four hours straight. Thus, they are usually at their sleepiest their first night on the night shift.

The immediate and predictable consequence of sleeping poorly is increased sleepiness during waking hours, particularly during night shifts. Surveys suggest that three fourths of all shift workers feel sleepy on every night shift. Of fifteen thousand shift workers surveyed by Circadian Technologies, Inc., 63 percent reported nodding off on the night shift at least once in the preceding week. Only about one quarter reported doing so on the day shift or the evening shift. The implications of these figures for public safety are staggering; it is miraculous that major disasters do not happen every night.

Irregular work schedules further worsen sleep. Under Federal Railroad Administration rules, railroaders may work shifts up to eleven hours and fifty-nine minutes before receiving eight hours off. They can be called back right after those eight hours. "So every day they're moving back four hours," according to Marty Klein of SynchroTech, a shift work consulting firm in Lincoln, Nebraska. "Crews may be notified only ninety minutes before their train is scheduled to come in," he said. "People get caught short on sleep all the time."

Whether many years of shift work will disturb sleep permanently—perhaps by instilling poor sleep habits and low sleep expectations—is still being debated. One study provides reassurance that that may not be the case. Wilse Webb compared fifty- to sixty-year-old female nurses with and without extensive night-shift experience. The nurses currently on the night shift slept poorly as expected, but the ones on the day shift, regardless of prior experience, said they usually slept satisfactorily and almost always felt rested when they awakened.

Gastrointestinal symptoms are more frequent. Twice as many shift workers as dayworkers develop gastrointestinal problems. "Nothing tastes good anymore," is a common complaint. For most people, part of the pleasure of eating comes from the environment in which they eat and the company with whom they share the meal. Night workers often have access only to cold or microwavable food from vending machines, a poor substitute for the hot meals served during the daytime in the typical company cafeteria. This environment deprives night workers of the natural camaraderie dayworkers share. Rather than eating only during a defined mealtime, many night workers snack throughout their shift, often right at their workstations, a habit that throws off balance normal intestinal enzyme activity.

Shift workers consume more caffeine and alcohol than people who work only during the daytime; both can be viewed as self-medication—caffeine fosters alertness and alcohol induces sleep. Some report their daily intake of coffee in pots, not cups. Nearly half of all shift workers smoke, about twice the rate of the general population. All these factors contribute to what is called "graveyard gut": about half report problems with diarrhea, and one third have constipation, particularly on the first few days after a shift change. By five years on the job, shift workers develop twice the incidence of peptic-ulcer disease as dayworkers.

- ***Heart disease rates go up.*** After five years on the job, shift workers are twice as likely to suffer heart disease and heart attacks as dayworkers of the same age and sex. After fifteen years, the incidence is three times as high. Shift workers have higher cholesterol levels than dayworkers, and they are also more likely to have high blood pressure. Here again, higher rates of smoking are thought to play a role.

People in high-stress jobs, such as air-traffic controllers, are at especially high risk for diseases of the cardiovascular and digestive systems. Sidney Cobb and Robert Rose of the University of Michigan found that male air-traffic controllers had four times the frequency of high blood pressure and twice the frequency of peptic ulcers expected for men their age.

- ***Female shift workers may have more menstrual-cycle irregularities and more complications of pregnancy.*** Some 20 percent of three thousand female workers surveyed by Circadian Technologies blamed shift work for making their periods occur more or less often and their flow more scanty or heavy. Many female flight attendants and nurses say that menstrual problems are so common that they are taken for granted.

In a study of nearly eight hundred female nurses on all shifts, Kathryn Lee of the University of California at San Francisco found that the nurses who worked nights or who rotated shifts were most likely to suffer menstrual-cycle irregularities. Although nurses as a group are generally healthy, the night and rotating shift nurses suffered more health problems in general than those on other shifts.

A few studies suggest that women shift workers have a slightly increased risk of having a miscarriage or preterm birth, and of having babies whose birth weight is lower than average.

- ***Older workers have more work-related health problems.*** As shift workers age, their pay increases, they often move to better housing, their children grow up and leave home, which means there is less noise to interfere with sleep. They have more rather than fewer sleep problems, however, because sleep deteriorates with age. Other health problems increase with age, and can be expected to become more widespread as the workforce running the nation's factories grays. Most shift workers are currently between ages thirty and forty-five, but among the nation's steelworkers, the average age jumped from forty-one to forty-seven between 1984 and 1989. At General Motors Corporation, the average age rose from thirty-six to forty-three in the last decade.

- ***Accidents and injuries become more frequent.*** Shift workers have auto accidents twice as often as people who work from nine to five. One in four shift workers reports having fallen asleep while driving to and from work.

  A study of medication prescribing errors in a teaching hospital found that the highest incidence of such errors, about four per thousand orders, occurred between noon and 4 P.M., the sleepiest hours of the waking day. Close behind were errors made between midnight and 4 A.M. Medication mistakes made during these hours put patients at higher risk than those made at other times. These include potentially fatal overdoses or underdoses. To minimize such errors, some hospitals have switched to unit-dose systems, in which the pharmacy supplies drugs in single doses, rather than in larger quantities to be doled out as prescribed.

  About one third of the night and rotating-shift nurses in Kathryn Lee's study acknowledged making errors. By contrast, only about one sixth of the permanent day or evening nurses did so. At night, the hospital corridors are quieter, lights are turned down, fewer doctors are on the floor, and patients are sleeping or trying to sleep. It is harder for nurses to stay alert then, too. "Nurses do fine on the night shift with intensive-care tasks such as setting up an intravenous solution," Lee reported. "But they are more apt to forget that it is hanging and about to run out." In one tragic case cited in a 1991 congressional report, *Biological Rhythms: Implications for the Worker*, a nurse stuck herself with a needle in the twelfth hour of a twelve-hour shift. The needle contained HIV-infected blood, and the nurse became HIV-positive.

  The 1992 report of the National Commission on Sleep Disorders Research states that between the hours of 2 and 8 A.M., a night worker, even one who has slept reasonably well the preceding day, is no more alert than a dayworker who has had only four hours of sleep for two nights in a row.

- ***Use of some drugs increases.*** Shift workers use more alcohol and over-the-counter drugs than dayworkers. Men who work variable shifts are more likely than dayworkers to consume four or more alcoholic drinks a day, and to use digestive aids on a regular basis. Women who work variable shifts tend to use more sleeping pills and tranquilizers than women who work straight shifts, and they also drink more. Compared with dayworkers, shift workers of both sexes are more frequent users of laxatives, sleeping pills, painkillers, and cough medicine.

- ***Emotional well-being may suffer.*** Shift workers report lower self-esteem than their dayworking peers. Perhaps this is a reflection of the stresses of coping with family and social demands in a world that regards daywork as the norm, and work at other hours as aberrant. Changing the timing of one's sleep, meals, and other biologic rhythms is stressful in itself, and may trigger depression in individuals who cope well under ordinary circumstances.

  Shift workers often describe themselves as "second-class citizens," an image reinforced by popular culture, where they are usually represented as faceless. Work around the clock poses unusual stresses, challenges, and potential life and death crises in which individuals need to think fast and creatively. These themes are the staples of television sit-coms. While various TV shows have focused on police, reporters, doctors, and others with nontraditional hours, the impact of living and working those hours has generally been given short shrift.

  "Nighttime on television is a time of danger and wackiness," according to David Zurawik, television critic for the Baltimore *Sun*. Shows involving emergency or rescue workers often take place at night, using darkness and flashing lights to heighten the dramatic impact. A show called *Midnight Caller* featured a policeman turned late-night talk-show host who took to the streets as a sort of Lone Ranger in response to calls for help. *Night Court* showed bizarre behavior in the evening. "People who work at night or in blue-collar jobs get no reinforcement from TV," Zurawik said. "On television, people talk about being promoted to the day shift. And when ratings go up, characters move up the socioeconomic scale fast."

  When the series *Roseanne* began, its star played a factory worker. The program had the potential to explore confrontational positions with management, but Roseanne soon was laid off and domestic crises became the show's central theme. Perhaps the best-known blue-collar worker on television is a cartoon character, Homer Simpson of *The Simpsons*, a safety inspector at a nuclear power plant. Simpson is presented as an oafish bungler, though, nobody's role model.

- ***Life expectancy may be shortened.*** Given the increased rates of health problems in shift workers, one might conjecture that their life expectancy would be reduced. There are no data showing that this is true, however. People with serious health problems may drop out of shift work early, and in any case, relatively few of them spend their entire careers doing shift work.

  Insects and laboratory animals subjected to a six-hour shift of their light/dark cycle, a schedule similar to that of rotating-shift workers, had a 20

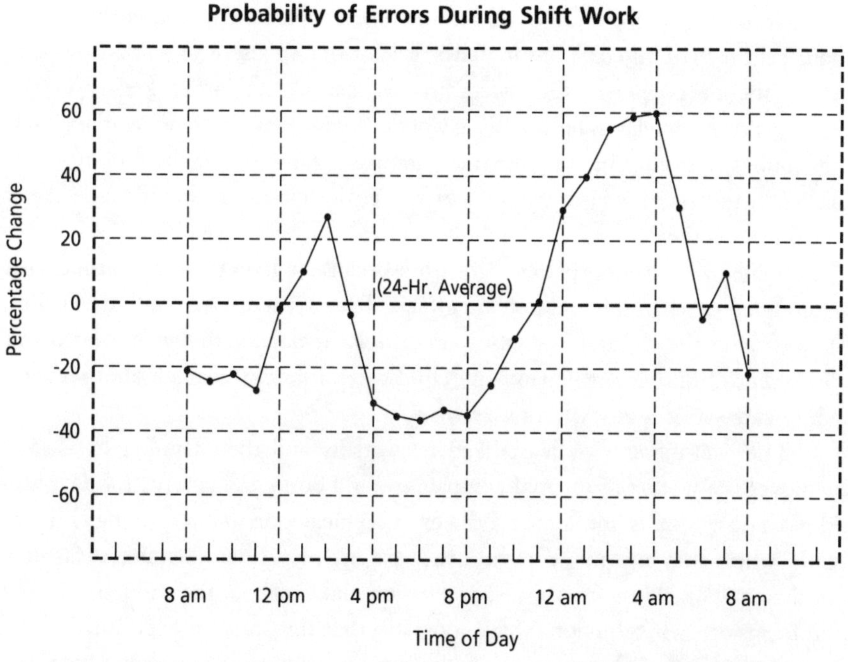

12:04 A.M. *Exxon Valdez*
12:40 A.M. Bhopal
1:23 A.M. Chernobyl
4:00 A.M. Three Mile Island

The likelihood of errors by shift workers varies over the twenty-four-hour day. At 4 A.M., the average shift worker is about 60 percent more likely to make a mistake in manual dexterity or to make a faulty decision. From 8 to 11 A.M., he or she is 20 to 25 percent less likely to do so. The graph, compiled by chronobiologist Charles Ehret, combines results of studies at several centers over thirty years, and it includes such errors as falling asleep at the wheel and delays in answering telephones.

percent shorter life span than those on a normal day/night schedule. People do not, of course, lead lives as regimented as laboratory animals; whether this is a plus or a minus remains to be seen.

IMPACT ON FAMILY LIFE

Most people prefer to go to work or school in the daytime. They view the evening as the time for the main family meal and family activities, even if that means just sitting together in front of the television. They prefer the night for sex and sleep.

Nonetheless, one third of all two-income couples with children under age fourteen in the United States includes at least one spouse who works a nonday shift. In one couple in ten, according to sociologist Harriet Presser of the University of Maryland, both spouses work different hours with no overlap. That maximizes the time that at least one parent can take care of the children, but leaves little time for both parents, or the entire family, to be at home—and awake—together.

Some single parents prefer night work because it gives them more time with their families when the children are awake. They say it is easier to find someone to look after the children at night than during the day. Nationwide, only about 5 percent of businesses offer on-site child care; around-the-clock and weekend child-care services are still a rarity.

The evening shift is least liked by workers and their families because it removes the worker as a social companion, and often as a parent, for days at a time. The worker is not home for the evening meal, and may see young children only when they are in bed; the other parent may resent having to assume primary responsibility for day-to-day decision making about the children's school problems or their behavior. Many complain that the children exploit the situation and attempt behavior they wouldn't pursue if both parents were present.

When a worker is on the night shift, the spouse who stays at home often complains of loneliness, and when it is the husband who is away at night, both spouses voice concern for the family's safety. Weekend work may interfere with both social life and parenting. Saturday night is the preferred night off; on a typical twenty-eight-day rotation, however, a worker will have only one Saturday night free per month. One's sex life demands flexibility: daytime lovemaking is subject to many interruptions and distractions. In one study of shift workers, nearly half called life on the night shift "sexless." Only one in ten made this complaint about other shifts.

Children of shift workers, like those of divorced parents, often take on family chores that the absent parent might otherwise do. Sometimes the burden is too onerous: nine-year-olds in some families are expected to baby-sit and make dinner. The children's school performance does not seem to suffer, however. The National Child Development Study in England, which followed sixteen thousand children since their births in 1958, has found that shift workers' children do as well in school, and are as well-adjusted emotionally, as dayworkers' children.

"Most shift workers carry a certain amount of guilt about not being accessible to their families, so they put their own needs aside," said Susan Koen of MATRICES Consultants, Inc., a shift-work consulting firm in Portland, Maine. "The family must make adjustments regardless of whether the shift worker is male or female," she asserted. "The home environment has to support the individual employee." The prior experience of the non–shift-working spouse, Koen said, influences a family's ability to cope. Spouses who themselves were reared in a home with a shift worker not unexpectedly find it easier to adapt.

Even in dual-career households, women do the bulk of the housework. Housework amounts to an unpaid second job for female workers, according to sociologist Arlie Hochschild. Women spend about fifteen hours a week more than men on housework, she reported in her book *The Second Shift*; they do two thirds of the daily jobs. "Most women cook dinner and most men change the oil in the family car," she said. Women perform more of the repetitive tasks; men do most of the fun chores. More women than men feed and bathe the children. More men than women take the children to the park, the zoo, and the movies. Women workers in general are more likely than men to report being "overtired, sick and 'emotionally drained,' " Hochschild said, adding, "these women talked about sleep the way a hungry person talks about food."

In her study of eight hundred nurses, Kathryn Lee found that some of them elected a permanent night shift, so that their spouse or partner could work during the day and be home at night to care for the children. "But if something went wrong at school," Lee said, "it was the woman who was called at home, and awakened to come and get her child. The women stayed in traditional roles with regard to parenting responsibilities."

Female shift workers who have children at home average eighty minutes less sleep every day than their dayworking peers. In a study of night nurses, Timothy Monk and his colleagues found even part-timers who had children living at home slept less than full-timers with no children at home.

The shift worker's demand for quiet during the day may be a source of family friction. A spouse and children may have a hard time restricting their activities or may resent doing so. In some cases, a vicious circle develops: the fatigued worker becomes more irritable, and the stress and strain make it harder to sleep.

Statistics show that more shift workers are separated or divorced than dayworkers. Such figures need to be interpreted cautiously: shift work may tear marriages asunder, or it may be more attractive to single people because of higher pay, more daytime hours off, or other features.

IMPACT ON SOCIAL LIFE

Because of their schedules, shift workers join fewer organizations than dayworkers, and they are less active in political parties, parent-teacher associations, and other community groups. Since organizations tend to hold meetings on evenings and weekends, shift workers attend fewer meetings of those they do join, and they are less likely to hold office.

Friendships suffer, too. Friends of shift workers may stop calling because they frequently find the worker asleep, or because work schedules and social events too often conflict. Shift workers miss more of the activities that serve as the basis for small talk in social relationships, such as sporting events or television shows. It is not surprising that many shift workers say most of their friends are other shift workers. On the plus side, many shift workers report that those friendships are unusually close.

Shift work has some advantages: significant amounts of time are freed up during the day during which shift workers can shop when stores are less crowded; they can also golf, garden, and utilize the many services that are offered only in the daytime. People who work permanent evening or night shifts can go to classes during the day or even take a second job. On the night shift, workers generally have more autonomy, since fewer managers are present. People who like tinkering with the car, sewing, or fishing often prefer shift work because it gives them more time for their hobbies. Indeed, one study showed that they pursued solitary hobbies more avidly than dayworkers.

"The social and domestic factors are at least as important in a person's ability to cope with shift work as the biological ones," Monk said. In company towns, where shift work is widespread and factored into social and community life, people seem to tolerate it better.

SOME TOLERATE IT BETTER THAN OTHERS

About 10 percent of shift workers enjoy the frequent schedule changes, and 60 percent tolerate them fairly well. Still, that leaves a sizable minority—one out of every three shift workers—for whom this type of life is a hardship. Certain individual characteristics reportedly predict one's ease of adjustment to shift work, and it has been proposed that hiring decisions might be based on screening workers for these traits. Not surprisingly, this suggestion has raised specters of possible discrimination predicated on genetic differences. Although Soviet scientists have used profiles of temperature and other biologic rhythms to screen cosmonaut candidates, such information at present benefits mainly the individual worker, who can use it in making career decisions. Research has shown that:

- ***Owls do better than larks.*** Owls are more alert at night, and are able to sleep later in the daytime. These characteristics are thought to make it easier to vary work schedules. People with the delayed sleep-phase syndrome—whose owlish tendencies are so extreme that they would have difficulty accommodating to daywork—may be able to turn their disability to an advantage by electing permanent night work.

  Nancy Ann Jenkins Hilliker and her colleagues at Deaconess Hospital in St. Louis found that larks who worked at night were sleepier than owls, and they slept more poorly during the day. Larks woke up more often than owls, and complained more about the quality of their sleep.

- ***Youth is a big plus.*** Young adults adapt more easily to evening and night shifts than workers aged forty-five and older.

- ***Flexible sleep habits help.*** People who fall asleep easily, in any setting, get more sleep than those who must have a quiet, dark place. People who are rigid sleepers usually find it hard to go to sleep early, or to sleep late, even when tired.

- ***Short sleepers do better than long sleepers.*** People who need less than six hours of sleep can move their sleep time around more easily than those who need nine hours or more.

- ***Wide-ranging body temperature rhythms, and longer free-running clocks,***

predict ease of adjustment. Daily highs and lows in body temperature vary in some people by more than two degrees, and in others, by less than one degree. Some people have a free-running clock that runs close to twenty-five hours; while in other people, it runs close to twenty-four hours. Those who have a wider range in temperature rhythms, and a longer free-running clock—and are also more likely to be in their twenties than in their forties or older—adapt more easily to changing schedules.

Constructing a temperature profile (see "Self-test: Alertness and Temperature Cycles," page 47) may provide a sense of whether one is likely to find it easier or harder to adapt to shift work.

- ***Extroverts cope better than introverts.*** People who are outgoing seem to adapt faster to shift work. Paradoxically, introverts often prefer shift work, especially nightwork, because it gives them more independence.

- ***Commitment to shift work fosters adaptation.*** People who enjoy working at unusual hours may pay more attention to getting enough sleep, eating regular meals, and scheduling recreation and other off-hour activities necessary for a balanced life. Moonlighters, who often work longer, and stranger, hours than most others, complain less about shift work; they may be the most able to adapt.

STRATEGIES TO IMPROVE SHIFT WORK

The growing appreciation of both the health problems suffered by individual shift workers, and the potentially dire impact of shift-work schedules on society has prompted some academics to hang up their white coats temporarily and put on steel-toed boots. The design of schedules to improve worker well-being and job satisfaction, while meeting the needs of a particular industry, has been a learning experience all around.

The story begins in 1980. "I have a hundred and thirty men who can't sleep. Can you help?" Preston Richey, plant manager at the Great Salt Lake Minerals and Chemicals Corporation in Ogden, Utah, had phoned chronobiologist Charles Czeisler after reading a newspaper article describing research by Czeisler and his colleagues, and its possible application to shift work.

Many of Richey's workers rotated shifts weekly, going backward around the clock. They complained about insomnia far more than their peers who were on permanent shifts. Nearly one third admitted having fallen asleep on the job at least once in the previous three months. One quarter said they rotated shifts so often that it was impossible to adjust their sleep schedule before they had to rotate again.

Czeisler and his fellow researchers, Richard Coleman, then also at Stanford University, and Martin Moore-Ede, then at Harvard, knew that a counterclockwise rotation like that at the Utah plant opposed the body's natural preference, which is to lengthen the day. Although the eight-hour rotation backward compressed the day, the body perceived it as a sixteen-hour switch forward. Consequently, adaptation to the counterclockwise rotation took much longer, and caused more sleep problems, than a clockwise rotation presumably would have done.

No one had compared a forward and a backward rotation at a real worksite. This was the first such study. The researchers devised a schedule that reversed the direction of the rotation. Additionally, they recommended that workers rotate at a slower pace. Eighty-five workers at the potash harvesting plant participated in the study; some continued to rotate at weekly intervals, while the rest stayed on their shift for three weeks.

Nine months after the schedule changes, nearly all the workers preferred the clockwise rotation. Among those who changed shifts every three weeks, complaints that the schedule changed too often fell markedly; personnel turnover also dropped.

Plant production rose by 20 percent the first year on the new schedule, and by 30 percent the second year. It continued to rise the third year, even though problems with the evaporation ponds where the potash was harvested forced an 11 percent cut in the workforce. *Business Week* hailed the experiment as "the study that convinced many companies that chronobiology was more than an academic discipline."

A NEW INDUSTRY: SHIFT-WORK CONSULTANTS

Since that study, Czeisler, Coleman, and Moore-Ede have all launched organizations to apply their academic expertise to the workplace. Czeisler, now at Harvard, serves as scientific adviser to ShiftWork Systems, Inc. Coleman runs the Coleman Consulting Group, Inc., and Moore-Ede heads Circadian Technologies, Inc.

These and other consulting firms (see nationwide directory, page 265) provide doctoring for ailing companies. Staffers spend time at a plant, observing, for example, the types of tasks performed, and characteristics of the work environment, such as lighting and noise. They interview workers about their schedule preferences, sleep habits, health, and the impact of their schedules on family life. They then develop schedules aiming at reducing problems, and they offer educational programs to teach workers and their families improved coping strategies.

Some of the consulting firms also conduct laboratory and field studies. Circadian Technologies' research center includes a mock-up of an industrial control room with a process-control simulator that can be programmed to run automated machinery such as is used around the clock in nuclear power plants and in chemical, paper, and textile industries. Shift workers participating in studies live in adjacent apartments and go to "work" on their designated shifts in the laboratory, where they sit at the control panel and perform their usual tasks. When something goes wrong, they must respond to alarms and attempt to set matters right. Computers keep tabs on what they do, and when they do it. The workers periodically take a battery of performance tests, and the researchers also monitor their sleep, temperature, and other biologic rhythms.

Czeisler helped the Philadelphia Police Department and its union, the Fraternal Order of Police, to develop a pilot program to improve the police officers' rotation schedule. Czeisler redesigned the schedule in the mid-1980s so that workers changed shifts every eighteen days instead of every eight days, rotating clockwise instead of counterclockwise as they had before. The number of consecutive days worked was reduced from six to four to give the officers a chance to recover lost sleep.

After an eleven-month trial in one police district, the improvements were dramatic. Automobile accidents on the job were 40 percent less per mile driven than in the previous two years. Sick time fell by 23 percent. The officers also reported less dozing on the job, and lower use of alcohol and sleeping pills. Family satisfaction with the new work schedule jumped fivefold. Although the officers had no more time to spend with their families, their families said the quality of time that they did spend was much better.

Despite these clear advantages, the program was not implemented citywide. It became a victim of a failed labor-negotiation process. The new schedule called for fewer, though slightly longer days, and that proved to be a sticking point that arbitration was not able to resolve.

THE ART OF SCHEDULE DESIGN

There are no simple or universal solutions as to how best to design shift work schedules, particularly those that run continuously. These are the main issues:

Direction of Rotation

Chronobiology studies suggest that work shifts should rotate forward around the clock. Certainly, body rhythms become resynchronized faster on this type of schedule, and there may be other health benefits: a Swedish study examined policemen with high blood pressure who worked a counterclockwise schedule. Half of them were switched to a clockwise schedule; by the end of the first month on the new rotation, their blood pressure was much improved.

Although a forward rotation is preferable from a circadian point of view, a backward rotation has some positive features, too. A backward rotation gives workers more time off between rotations, and thus may enable them to get more sleep. Whether they actually experience less sleep deprivation on a counterclockwise schedule has not yet been determined. In the United States, about half of the country's workers currently rotate in each direction.

Speed of Rotation

There is little consensus on the number of days that workers should spend on each shift. In the United States, weekly rotations are most commonly used; they are known as medium rotas. European countries more often use shift schedules that rotate every few days, known as rapid rotas, continental rotas, or metropolitan rotas. On such schedules, a worker might work two-day shifts, then two evenings, then three nights, and then have two days off. Slow rotas are schedules that change after two or more weeks on the job.

Support for one type of rota over the others depends on the goals of whoever designs the schedules. Those who favor adaptation advocate slow rotas, preferably shifts lasting three weeks or more. Ideally, workers would maintain the same schedules on their days off, but human behavior rarely meets this goal: on days off, even permanent night workers usually revert to daytime schedules.

Those arguing against adaptation favor rapid rotas, on which body clocks remain relatively day-oriented, particularly the slow-to-change temperature rhythm that has a major influence on performance. Workers on rapid rotas can sleep at night about 75 percent of the time; of course, when they work at night, their performance will fall off. Further, after the night shift, their sleep is so disturbed that workers may end up using days off to catch up.

There is no question that a weekly rotation is harder than a more rapid rota. A backward weekly rotation is more difficult still. "It is like driving down the freeway at a hundred miles per hour in the wrong direction," Richard Coleman said. "It won't help much to put on a seatbelt and slow down a little."

Unpredictable or irregular schedules are worst of all: body clocks stay perpetually out-of-sync.

In one study, Martin Moore-Ede and his colleagues compared job alertness in workers on three types of schedules: when they rotated every seven days, counterclockwise, or every two days, clockwise, they rated themselves in a "low alert" state at work more than one third of the time. However, when workers rotated every twenty-one days, clockwise, they rated themselves as "low alert" only 6 percent of the time.

The rotation speed that is "best" may depend on the tasks that have to be performed. Timothy Monk studied people who had to enter coded information into the computer that controlled their large chemical plant. Because they worked a rapidly rotating twelve-hour schedule, they stayed oriented to the daytime. For a month, the computer automatically logged every mistake that they made. They actually made fewer errors on the night shift, although they did as much work as on the day shift. The particular task that they had to do required short-term memory, an ability that remains high at night, and their lack of adaptation to the night shift was not a problem. For other types of jobs, slow rotas might be better. These include tasks that require vigilance in monitoring computers or dials, some of which have life-or-death consequences, such as air-traffic control. By promoting adaptation, slow rotas would reduce the tendency to make errors of inattention at night.

Length of Workday

Hours of work are getting a new look because of chronobiology research. The popularity of twelve-hour shifts has prompted concern about the impact of fatigue from long hours, particularly on the night shift (as well as longer continuous exposure to such hazards as workplace chemicals and noise, not biologic clock issues, but certainly ones that may undermine health and job performance).

In one study, Roger Rosa and his colleagues at the National Institute for Occupational Safety and Health found that people who worked twelve-hour shifts did worse on several tests of alertness, even after seven months, than those on eight-hour shifts. There was no test on which performance improved on a twelve-hour schedule. The workers' actual performance on the job, however, which involved filling out log books and noting unusual events, was comparable on both the eight-hour and the twelve-hour shifts. Such studies show the difficulty of translating findings from the laboratory to the workplace, and the need to devise meaningful tests.

Further, even though increased hours have classically been linked with increased fatigue, a schedule could be devised in which workers would spend more hours in a row on the job, but work mainly during their peak alertness daytime hours. Truckers, for example, are currently supposed to take eight hours off after ten hours on the road, a requirement that often puts them back behind the wheel at night. If they regularly started at 6 A.M., even if they drove for fourteen hours, they could keep their body clocks in sync and sleep at night, and they might have fewer accidents.

The type of task to be performed needs to be factored into the amount of time that workers spend on the job. Long hours are likely to lead to performance decrements in jobs that require intense concentration, such as performing surgery; or are monotonous, like watching a video monitor; or involve heavy physical activity—firefighting, for example.

Needs of Employees and Employers

The needs of employers and their workers vary widely, even in the same field, and sometimes even in the same company. For example, a Fortune 500 company in the rubber and plastics industry decided to change its work schedule from five days to seven days a week around the clock in several of its plants. The company sought help in designing a schedule from Donald Tepas of the University of Connecticut and his colleagues at the Illinois Institute of Technology and Work Systems Research.

The researchers surveyed more than two thousand workers in four different plants. Even though all of the plants performed similar work, nearly 15 percent of the workers in one plant held second jobs, while fewer than 5 percent of those in another plant did. For the first plant, a twelve-hour shift system would have been a bad idea, Tepas asserted, because the moonlighting workers would have become too fatigued. In the second plant, a twelve-hour shift system would have been more acceptable.

Computer-assisted time scheduling facilitates the design of schedules that meet both employer and employee needs. Much like computerized airline-reservations systems, such programs can produce schedules several weeks in advance, considering organizational, health, and legal requirements, as well as worker preferences. The computer could match up workers with complementary schedule requests or needs so they could substitute for one another, and even permit requests for a change of schedule when workers were fatigued, thus potentially cutting back on costly errors. As an added bonus, this program could reduce personal supervision of requests. That would relieve both workers and supervisors from hassles over work hours, and improve communications and morale.

Some employees charge that schedules often fail to consider their lives off the job. Or management policies may sabotage even reasonable schedules. "Some companies compensate for understaffing with overtime. That's particularly hard on families on weekends," according to SynchroTech's Marty Klein. "Workers and their families come to dread phone calls," he said. "A kid will answer a call and say, 'I don't know if my dad's home. I'll ask him.' Companies need to adopt the policy that the last person to call for overtime is the one who is off for the weekend," he asserted.

Workers in small isolated towns, who prefer long blocks of time off so that they can get out of town, will be willing to work more days in a row to achieve them. By contrast, those who work in areas with many leisure attractions may prefer to have fewer, but shorter, breaks. Some schedules provide three days off. The company might make those days Sunday, Monday, and Tuesday, although almost invariably, Saturday is the preferred day off. The company may say it is too much trouble for the accounting office to change the start of the workweek to accommodate that preference, although there may be six people in the accounting office and five hundred shift workers. The start of the workday may be a similarly loaded issue. A difference of an hour or two at the start of the day shift, for example, may enable workers to synchronize their schedules better with those of their families or avoid rush-hour traffic. Subtle practices, such as the time a holiday party or company picnic is scheduled, may affect morale; a company needs to schedule both daytime and nighttime events, so that all workers can attend. While increasing numbers of companies have on-site fitness centers, unfortunately, not all staff them for the evening and night shifts. (For coping strategies for shift workers, see the Appendix.)

There are plenty of payoffs for management in good schedule design. The Amax Coal Company in Indianapolis faced the choice of spending $14 million on new equipment or utilizing its existing equipment more fully with weekend work. Richard Coleman designed a schedule that rotated workers every twenty-eight days with a seven-day break between shift changes. Workers made more money on the new schedule, plant productivity rose 10 percent per labor-hour, and the company was able to shelve its plans for the far more costly capital improvements.

FROM THE LAB TO THE WORKPLACE

Improved schedule design is only one benefit of chronobiology research. Other studies focus on devices to monitor fatigue, redesign of the workplace, medications to improve sleep, and bright light to improve alertness.

Development of Fatigue Meters

The transportation industry, in particular, could benefit from devices designed not so much to police performance as to facilitate alertness. For truckers, such devices might monitor variability in the steering wheel, or check the roadway and sound an alarm if the truck strays across the white line. Instruments could be built into the console of the cab to warn drivers when their eyelids droop, or to alert them if they come too close to the vehicle in front of them.

Reengineering the Workplace

Factories used to be brighter, noisier, more crowded, and more uncomfortable. One has only to think of Charlie Chaplin's film *Modern Times*. Today, workers often sit in a comfortable chair, in a darkened room, hunkered over a computer screen. Chronobiologists have suggested reviving some former aspects of the worksite to promote alertness: turning up the lights, turning down the heat, opening windows, making floors vibrate and instruments beep, playing lively music, redesigning tasks so that workers will have to move around more, if only to get up to push a button on the opposite side of the room, and promoting interaction with fellow employees. The most recent area of research has focused on the impact of aromas; peppermint, for example, is alleged to have an alerting effect.

Sleeping Pills May Improve Daytime Sleep

Sleeping pills may help night workers to sleep a little longer during the day. They thus would be most useful on the first day or two after a shift change, before significant adjustment of body clocks has gotten under way. Unfortunately, sleeping pills do not improve alertness on the job, according to a study by James Walsh and his colleagues at Deaconess Hospital in St. Louis. Regardless of pill use and amount of sleep, for the first two or three nights after shifting from daywork to night work, people were profoundly sleepy between 2 and 7 A.M. With or without a sleeping pill, people in their twenties slept about a half hour longer than those in their thirties and older, the researchers found. The young adults also overcame their sleepiness a night or two sooner than the middle-aged group.

Drugs of potential value for shift workers are the same as those that may relieve jet lag. (For more information on these drugs, see pages 174–175.)

Lights May Boost Alertness

The potential of bright lights to shift body rhythms has direct application to the workplace. The challenge is to devise strategies that speed adaptation, with the least harm to sleep and mood, while taking into account the conflicting cues that workers receive from seeing daylight on the way to and from work.

In a study devised by Charmane Eastman of Rush-Presbyterian-St. Luke's Medical Center in Chicago, twenty-four young adults lived on a conventional daytime schedule with fixed bedtimes and waketimes for ten days. Then they stayed awake all night, and the next day began a new schedule on which they went to bed and got up exactly twelve hours later—a change in routine akin to one shift workers often must follow. They followed the new schedule for at least eight days, staying awake at night, and sleeping during the day, while they continued to live at home, and engaged in their normal activities. To prevent exposure to sunlight, they covered their bedroom windows with black plastic, stayed away from sunny windows, and wore dark goggles when outside. All received an equal dose of five thousand lux light, six hours for the first four nights, and three hours thereafter. But some participants obtained most of their light exposure *before* their body temperature low point occurred, while others got most of their light exposure *after* the temperature trough.

The temperature rhythms in most subjects shifted by about two hours a day. By contrast, those in most shift workers rarely shift at all. While rhythms theoretically should shift after any change in schedule, the dose of daylight shift workers receive on the way home generally blocks such changes.

In Eastman's study, rhythms in subjects who received most of their light exposure before their temperature minimum shifted forward around the clock, while rhythms in those who received most of their light exposure after their temperature minimum went backward. This difference has some important practical implications for light-savvy workers of the future: knowing the direction one's rhythms were shifting would enable one to anticipate highs and lows in performance, and adjust one's schedule accordingly. Alertness would be high at quitting time in night workers whose rhythms were shifting in a counterclockwise direction, for example, but low in those whose rhythms were moving clockwise; those in the latter group would be wise not to drive, but they should be able to count on falling asleep quickly if they go right home.

In another study using lights, Charles Czeisler and his colleagues asked eight young men who followed a conventional dayworker's schedule to work at night in their laboratory for a week. During their eight-hour night shifts, the men worked at a desk and were exposed to either bright light or ordinary room light.

They then returned home, receiving exposure to natural sunlight on the way. The ones who had been exposed to bright light were asked to stay in the dark at home from 9 A.M. to 5 P.M. They slept in bedrooms with light-blocking window coverings. Those receiving ordinary room light slept whenever they wished in rooms with the usual shades or curtains.

The men who both received bright light exposure at night and stayed in the dark during the day adapted to the night schedule within six days as indicated by changes in their temperature rhythms and hormone rhythms, and improvements in their alertness. By contrast, these rhythms remained unchanged six days later in the men who worked under ordinary lights.

In a third study, Drew Dawson and Scott Campbell of Cornell found that a one-shot light exposure—between midnight and 4 A.M. on the first night on the night shift—was enough to speed changes in body clocks, improve alertness on the job, and enhance daytime sleep. By the third night shift, workers' temperature rhythms had shifted by five to six hours, whereas those of workers not exposed to the lights had shifted only two to three hours. As the subjects in this study remained in the laboratory, they received no conflicting exposure to natural daylight.

"More work is needed to make these techniques more powerful, and more practical," Eastman said. Ideally, light treatment would let workers perform their normal duties on company time, and thus could be incorporated easily into any setting where workers stay put.

Indeed, bright lights proved a boon to writers and editors for CBS morning news shows who work a permanent night shift, midnight to 9 A.M., five days a week, tracking breaking stories. Typically, these workers reverse their schedules completely on weekends. Understandably, they have the most trouble battling fatigue on their Sunday night/Monday morning shift. During the week, like all night workers, they face a 4 A.M. ebb in alertness. Under the direction of Leslie Powers and her colleagues at the New York State Psychiatric Institute, eight newsroom workers received thirty minutes of bright light exposure each evening. "The lights are better than coffee," one writer claimed. The workers felt that they functioned at a higher level throughout the night. The lights held off the predictable drop in alertness until 7 or 8 A.M., near quitting time. Daytime sleep improved, too.

Bright lights might provide a similar benefit for students, who often study at night. George Brainard of Jefferson University in Philadelphia and his colleagues found that sleep-deprived volunteers who worked under bright lights far outperformed their peers working under dim lights. This research implies that students, as well as anyone staying up late to meet a deadline, should work in the brightest place possible and not rely on a solitary desk lamp for lighting.

The design of workplace lighting to both improve workers' alertness on the job and their sleep during off-hours has become a business itself. ShiftWork Systems uses computers to first determine desirable light exposure schedules for individual employees and then to regulate the overhead lighting at the employees' work stations. At different times during a shift, the lights might range from being about twice as bright as a typical room to the intensity of outdoor light on a cloudy day.

Both Eastman and her colleagues and Czeisler and his colleagues have devised regimens for the astronauts on more than a dozen space shuttle missions, using a combination of sleep at prescribed hours, appropriately timed exposure to bright lights, and avoidance of sunlight in the week before the launch. One news photo shows the astronauts on the launchpad, suited-up for space, but looking rather incongruously like ordinary tourists, as they are all wearing dark sunglasses.

SUMMING UP

Chronobiology studies have helped bring the plight of the shift worker into public consciousness, and have fostered efforts to improve the lives of shift workers and their families. Good schedules take into account the direction and speed of the rotation, the length of the workday, the personal needs of employees and their families, and the operational needs of the employer. Improved schedules offer numerous advantages to management by reducing errors on the job and employee turnover, and often by increasing productivity as well. The benefits for public health and safety are large, and the risks of inaction daunting. So why does the United States lack a national shift-work policy?

CHAPTER 11

A CALL FOR PUBLIC POLICY CHANGE

IN THE UNITED STATES, circadian principles have not been incorporated into national policy. Every single federal regulation relating to hours of work limits hours on the job and requires rest periods, but not one addresses the time of day that rest periods are to be taken. All were written without our present-day understanding of biologic clocks. No laws acknowledge that *when* people sleep may be even more crucial than *how long* they sleep in determining performance and well-being on the job and off.

Many other nations are far ahead of the United States in this regard. France mandates that workers who have performed night work continuously for five years, or for ten years altogether, receive top priority for openings on the day shift. France is also unique in providing night workers with extra pay to lightproof and soundproof their houses, so as to improve daytime sleep, and it has addressed quality-of-life issues by arranging for certain television programs to be rebroadcast when night workers can view them. A 1991 report, *Biological Rhythms: Implications for the Worker*, prepared for the U.S. Congress by its Office of Technology Assessment, compared work regulations in the United States with those in forty-eight other countries.

Austria requires all companies with more than fifty shift workers to have an on-site medical officer. It also requires that workers who perform heavy physical labor at night get extra days off. In Bolivia laws state that "night work is contrary to man's physiology, exhausting and dangerous to the health if practiced as a normal system," and night work is limited to a maximum of seven hours. In Japan the law requires that security guards and others on twenty-four-hour shifts be allowed a minimum of three hours for sleep during their shifts. Japanese companies provide quiet on-site sleeping rooms for regularly scheduled nap breaks.

In some countries, women are not permitted to work at night; this is not the case in the United States, where such restrictions are seen as discriminatory. The only restrictions on night work in the United States apply to children under age eighteen. Most other countries also restrict night work by young people. No country currently regulates whether shifts rotate forward or backward.

In November 1993 the twelve nations that make up the European Union voted to require member countries to enact laws by November 30, 1996, to limit the work week to forty-eight hours and night shifts to eight hours, and to require at least eleven consecutive hours of rest between shifts. The member countries include Belgium, Denmark, France, Germany, Greece, Ireland, Italy, Luxembourg, Portugal, Spain, the Netherlands, and the United Kingdom. The work hours directive is part of the European Union's *Charter of Basic Social Rights for Workers*.

In the United States, responsibility for hours of work is parceled out to an assortment of federal and state agencies. Some industries are more highly regulated than others, usually depending on the potential of the job to cause harm to the public at large or to the individual employee. Thus, transportation has many rules, while agriculture has few. The U.S. Department of Transportation regulates trains, planes, trucks, and ships. The Nuclear Regulatory Commission oversees workers in nuclear power plants. The Mine Safety and Health Administration covers all coal and other miners. The federal Occupational Safety and Health Administration has issued numerous standards to protect employees from dangers in the workplace, such as exposure to AIDS or carcinogens; none of its standards, however, currently deals with work scheduling.

The National Transportation Safety Board (NTSB), although it lacks regulatory power (it is an independent federal agency charged with accident investigation), routinely reviews the seventy-two-hours before an accident to evaluate the role of sleepiness in the events. Since 1972, the NTSB has issued more than three dozen recommendations to federal and state agencies, businesses, and associations regarding fatigue, time on duty, and hours of service. Unfortunately, these are only recommendations. They have no teeth.

REGULATIONS CONCERNING DOCTORS IN TRAINING

The death of a woman in a New York hospital in March 1984 focused attention on the grueling schedules worked by physicians in training, and prompted the nation's first laws regulating doctors' work hours.

Feverish and agitated, eighteen-year-old Libby Zion had been brought by her parents to the emergency room of New York Hospital at 11:30 P.M. Seven hours later, after her temperature had soared to 108°F, she died of heart failure. Her father, Sidney Zion, a lawyer and writer for *The New York Times*, charged that neglect by overworked doctors had caused her death. The residents who evaluated the young woman after her 2 A.M. admission to the hospital had been on duty continuously for the previous eighteen hours.

At the time the death occurred, residents commonly worked shifts of thirty-six hours, and weeks of one hundred hours or more. First-year residents reportedly averaged only two and one half hours of sleep per day. While both of the residents involved in the Zion case were cleared of negligence charges, New York legislators passed a law that went into effect July 1, 1989, limiting doctors in training to shifts of no more than twelve hours in the emergency room, twenty-four hours when providing nonemergency care, and workweeks of no longer than eighty hours, including moonlighting. The cost of hiring additional residents and other workers to provide the same level of service as existed before the law went into effect was estimated to exceed $200 million annually.

While New York is the only state so far to mandate limits on residents' work hours, many of the nation's hospitals have voluntarily adopted a limit of eighty hours of work per week. At the Johns Hopkins Hospital in Baltimore, Maryland, for example, neurosurgery residents are now on call every fourth night; a few years ago, they were on call every other night. And there is new attention toward reducing the residents' burden by designing better schedules. Some hospitals now use "night float" systems, in which one resident works a night shift every night for a week or longer. That resident sleeps during the day to give other residents time to sleep during their on-call night.

LAWSUITS: AN IMPETUS FOR CHANGE

Lawsuits affirming the legal liability of employers for their sleepy employees may further spur radical changes in shift work schedules.

In one case, which received a great deal of publicity, a jury found the McDonald's Corporation liable in 1991 for having permitted an eighteen-year-old employee to drive home after working all night in a McDonald's restaurant in Oregon. The employee, Matthew Theurer, fell asleep at the wheel and hit another car head on. Theurer was killed; the driver of the car he struck suffered serious injuries. Theurer, a high school senior, had attended class the day before the accident, worked from 3:30 to 7:30 P.M. that night, and returned to work at midnight, working until 8:20 A.M. the following day before starting the nineteen-mile drive home. He had not slept at all in the twenty-four hours before the crash, and had complained at work of being tired. According to the jury, McDonald's knew, or should have known, that Theurer would be a danger to himself and others while operating a motor vehicle.

There have been several other "sleepy worker" cases in recent years, according to a report on the Theurer case in *The Wall Street Journal*, most involving employees who worked very long or consecutive night shifts. In most of these cases, the employee had reported the fatigue to a supervisor.

WAKE UP, AMERICA!

It is not just the workplace that needs changing, of course. And perhaps changes in the workplace would be the natural outcome of an increased understanding of biologic rhythms by the general public, by physicians, and by legislators. At present, most people know little about their inner clocks, particularly the major daily rhythms of alertness and sleepiness. They have small appreciation of both the power of sleep to assure alertness, and the power of sleepiness to sap performance. *Wake Up America*, the National Commission on Sleep Disorders Research's 1993 report to Congress, asserted that "a substantial number of Americans, perhaps the majority, are functionally handicapped by sleep deprivation on any given day."

Few people appreciate that sleepiness is cumulative, and that such measures as drinking coffee, splashing cold water in one's face, and turning on loud music serve merely to camouflage sleepiness, not to reduce it. People who will arrange for someone else to drive if they plan to consume alcohol, nonetheless will get behind the wheel of a car when sleepy. Sadly, few realize that sleepiness itself compromises the ability to recognize that one is sleepy. The Department of Transportation has estimated that some 200,000 automobile accidents occurring in the United States annually may be sleep related. One out of every five drivers surveyed admitted to having fallen asleep behind the wheel at least once. Most driver education courses, the sleep commission found, neglect instruction on the dangers of driving when sleepy.

Science courses at the primary, high school, and college levels seldom cover body rhythms. Education about sleep in the nation's medical schools, the sleep commission stated, is also woefully inadequate. In 1990, a survey for the commission determined that 37 of the 126 medical schools in the United States offered just one hour or less of classroom education about sleep during the first two years, the years of preclinical instruction. Fewer than one dozen schools offered intensive clerkships or specialized elective training in the field.

This helps explain why the nation's physicians know little about sleep and biologic rhythms. The commission examined the records of patient visits to nine family-practice clinics in 1989–1990. Of the medical records for more than 50,000 visits, only 123 mentioned a sleep complaint. That is quite astonishing, considering that sleep problems are virtually universal among the one fifth of the nation's workers who work night or rotating shifts, and that one in three Americans reports suffering from insomnia. Questions about problems staying awake, or problems falling asleep and staying asleep, are not yet part of the standard medical history.

Additionally, physicians typically prescribe medications for most illnesses to be taken in equal doses throughout the day, a practice that fails to acknowledge that the body may function quite differently at different times of day. Nor does present-day drug testing take into account the impact of the time of day a drug is given on the drug's effectiveness or toxicity.

Considering the numbers of people affected by sleeping and waking disorders, federal support for research has been low. In fiscal year 1990, the National Institutes of Health and the Alcohol, Drug Abuse and Mental Health Administration together awarded just under $45 million for sleep-related research, whereas $1.67 billion was alloted for research on cancer, $1.58 billion for AIDS, and $672 million for heart disease. However, the research picture may improve soon: in 1993, Congress authorized a National Center for Research and Education on Sleep and Sleep Disorders, to be housed at the National Institutes of Health.

The establishment by Congress of a national commission to evaluate work/rest schedules and to recommend improvements might lead to legislation raising U.S. labor standards to those already in force in most other industrialized nations. America's workers surely are entitled to no less.

Changing public attitudes remains a daunting task in a society where many view sleep as a waste of time and workaholism as a positive trait.

At a workshop on hours of service sponsored by the U.S. Department of Transportation, after a day of listening to alarming reports on the pervasiveness and dangers of sleepiness, one member of the audience asked his colleagues: "Does anyone here work a forty-hour week? And if you did," he challenged, "would you admit it?"

APPENDICES

STRATEGIES FOR SELF-HELP

FOLLOWING IS A COMPENDIUM OF PRACTICAL information derived from the research reported in this book. It is a series of self-help guides, ways to harness the biologic clock to improve your life.

HOW TO TELL IF YOU ARE GETTING ENOUGH SLEEP

There is no specific amount of sleep that is right for everyone. Sleep needs are as individual as shoe sizes and cannot be changed. You can assess the adequacy of your own sleep by answering the following questions:

- ***Do you need an alarm clock?*** If your sleep needs are met, you will awaken spontaneously without needing an alarm clock. One caution: depressed people often awaken earlier than desired. Pay attention to your moods.

- ***Do you have trouble waking up?*** It is common, and perfectly normal, to feel sleepy on arising. Sleepiness should dissipate within a few minutes. If you need a cup of coffee to feel alert, you may not be getting enough sleep.

- ***Do you feel alert during the day?*** This is the prime test of whether you are getting enough sleep.

- ***Do you nod off easily at quiet times?*** Dozing in front of the television, or when a speaker drones on, is common, but people who have had enough sleep will feel restless, not sleepy, when bored. Warm rooms and dull situations do not cause sleepiness; they unmask it.

- ***Do you need to sleep late on weekends to catch up?*** This is akin to bingeing on food after several days of starving. You should not need to plan your days off around recovery sleep. A balanced sleep diet is better than fasting and feasting.

- ***Do you lie awake long after you turn the lights off?*** People who are ready usually fall asleep within thirty minutes. Perhaps you need less sleep than you think. Some people have bought the eight-hour myth but are really seven-hour sleepers.

FINE-TUNING SLEEP REQUIREMENTS

An hour or two of sleep more or less each day may not seem significant. But if you need it, more sleep might make you more alert and productive. If you feel fine without it, you could enjoy the extra time awake. It may be worth a trial-and-error experiment to find the amount of sleep that is right for you. Some suggestions:

- ***For just one night, set your alarm and sleep only half as long as usual.*** The next day, pay attention to how it feels to be sleepy.

- ***For a week or two, try to sleep the same number of hours every night.*** Keep your bedtime constant and get up at the same time every day. Then go to bed thirty minutes earlier or later for another couple of weeks, and see if you feel a difference in your sleepiness and alertness.

- ***Try taking a thirty- to sixty-minute nap every afternoon for a week or two.*** Do you feel reenergized for the rest of the day? Does napping make it harder to fall asleep later that night?

HOW TO COPE IF YOU EXPECT TO MISS SOME SLEEP

- ***Take a nap ahead of time.*** If possible, get some extra sleep for a few nights beforehand. If you bank extra sleep on, say, Sunday, you will feel more alert on Wednesday even if you miss some sleep on Monday and Tuesday. And on Tuesday if you take a nap before your late night, you will feel more alert the next day than you would if you wait to nap until sleepiness sets in.

- ***Get as much sleep as you can during your regular sleep time.*** Even a little sleep is better than none.

- ***If you regularly work long or irregular hours, build a nap into your schedule.***

HOW TO COPE AFTER MISSING SLEEP

- ***Socialize.*** Interaction with other people makes it easier to stay alert. If you work alone, make phone calls when you feel sleepiest.

- ***Put forth extra work effort.*** Motivation can override sleepiness for the short term.

- ***Exercise.*** An active body makes for an active brain. In the office, stand up and walk down the hall.

- ***Spend fifteen minutes outdoors right after arising.*** The alerting effects of light are greatest early in the day.

- ***Eat lightly.*** Avoid alcohol and carbohydrate foods.

- ***Take a midday nap, if possible.*** Do not nap after dinner.

- ***Drink caffeinated beverages judiciously.*** If you normally restrict your caffeine intake to only one or two drinks a day, a small amount can help you stay awake when you really have to. People who consume large amounts of caffeine become habituated to it and do not get much of a boost from consuming more. Light users of caffeine who are anticipating a night without sleep—college students studying late and doctors on call, for example—will find that a single cup of coffee, consumed between 10 and 11 P.M., revs alertness significantly for the rest of the night.

- ***Go to bed an hour or two earlier.*** This will keep you from carrying over the sleep debt to the next day.

ADVICE FOR STUDENTS

- ***Study material you need to retain, such as foreign-language vocabulary, in the afternoon when long-term memory is best.***

- ***Review your notes before an exam in the morning, when short-term memory is best.***

- ***If you must cram, go to sleep at the usual time, sleep four hours, get up, and take advantage of the normal rise in alertness in the early-morning hours.***

- ***At night, study in a brightly lighted room.*** Surrounding yourself with light will keep you more alert than simply using a desk lamp.

GUIDE TO GOOD SLEEP

Stabilizing sleep times synchronizes other body rhythms. People who follow established routines sleep better at night and feel sharper during the day.

- ***Establish regular hours.*** Try to get up at about the same time every day, no matter when you go to sleep. A regular time of arising is the single most effective way to keep body rhythms in tune.

- ***Create enjoyable presleep rituals.*** Watch the late news or a comedy show, or have a light snack. Doctors often advise patients to relax at bedtime; some people favor reading, while others enjoy hobbies, even doing exacting work. Some people can read or watch television in bed, but others find these practices are too alerting.

- ***Exercise for at least twenty minutes, three to six hours before bedtime.*** This practice will raise body temperature. The decline in temperature that occurs after exercise sends a strong signal to the brain that will help induce sleep a few hours later. Regular exercise also helps structure your day.

- ***Take a hot bath for twenty minutes two to three hours before bedtime.*** Keep adding hot water as necessary. This practice raises body temperature; the cooling down after the bath helps induce and deepen sleep.

- ***Avoid caffeine within four or five hours of bedtime.*** Even if it does not interfere with falling asleep, caffeine makes sleep more restless.

- ***Skip both alcohol and nicotine near bedtime.*** Alcohol may make you drowsy, but once that effect wears off, it will disrupt sleep. Nicotine is a stimulant.

- ***Keep your bedroom quiet.*** Mask any noise with the hum of a fan or air conditioner. You can get the effect of a white-noise machine by turning your radio to the static at the end of the FM band, or your television set to an empty channel.

- ***Keep your bedroom dark.*** If streetlights and headlights shine in, get room-darkening shades or curtains. Open them as soon as you get up in the morning. If your bedroom is secluded, sleep with the shades up. Sunlight will awaken you gradually and gently.

HELP FOR TROUBLED SLEEPERS

Many people view insomnia as a problem that must be endured. That is not so; there are many effective steps to take.

- ***Reserve your bed for sleep and sex only.*** Do not use it as a place to pay bills or talk on the phone; such activities will cause you to associate getting in bed with staying awake.

- ***If you are a bedtime worrier, set aside a half hour after dinner to jot down problems and possible solutions.*** Even if nothing can be done, you may find that writing a problem down helps you come to terms with it and lets you rest easier.

- ***To nap or not to nap?*** Some poor sleepers find that afternoon naps relieve their anxieties about not getting enough sleep, and make it easier to relax at night. Others find that daytime naps interfere with nighttime sleep. Try napping consistently for a week or so to determine its true impact.

- ***If you cannot fall asleep in ten or fifteen minutes, get up and go to another room.*** Engage in some quiet activity, such as reading or watching television, until you feel sleepy. Keep the lights dim and do not snack. Go back to bed only when you feel drowsy.

- ***If you cannot sleep, but it is too cold to get out of bed, try to focus your thoughts.*** One woman redesigns her garden in her head. Another takes a mental walk. Visualize your attention as a spotlight; shine it on a subject other than your concern about not sleeping.

- ***Try not to be distressed when you sleep poorly.*** You probably will sleep better the next night.

- ***Use sleeping pills cautiously.*** They may help for occasional bad nights or on trips away from home. But they are not designed to take every night, and indeed, they lose their effectiveness after about two weeks of nightly use.

- ***Keep a sleep diary.*** If the above steps do not improve your sleep, use the sleep log on page 232 to help identify trouble spots.

- ***If sleep problems persist, see your doctor.*** He or she may refer you to a sleep disorders center. For names of specialists in your area, write The National Sleep Foundation, 122 South Robertson Boulevard, Los Angeles, CA 90048-3208.

LIGHT: A USER'S GUIDE

- ***Consider the desired effect.*** Morning light may help you wake up earlier. Light in the evening may help you stay up later.

- ***If your bedroom is secluded, leave the windows uncovered.*** Let sunlight awaken you naturally. Going outside as soon as possible will give your biologic clock a strong start-the-day signal. This is particularly useful if you have trouble getting going in the morning, especially in the winter.

- ***If you suffer an afternoon slump, forgo a coffee break.*** Instead, take a walk outdoors.

- ***If you have trouble falling asleep at night, particularly in summer, stay indoors as much as possible after 4 P.M.*** Be sure to go outside in the morning.

SHIFT WORK COPING GUIDE

If your work schedule is other than nine to five, and particularly if you work at night, you can improve alertness on the job and your sleep, and adapt better to schedule changes by following these guidelines:

- ***Engage in some physical activity in the last few hours of your shift.*** Stand up frequently. Stretch. Walk around. Go outside if you can. If possible, save your breaks until your foggiest time. If you have a choice, do your most boring tasks early in the shift, and your more interesting ones later on. Try to drink some fruit juice or a decaffeinated beverage about an hour before the end of your shift. This will help make the last hour pass more quickly.

- ***Sleep as much as possible at night.*** When on a night shift, some people go to sleep in the morning right after work. Some sleep just before they go to work. Some split their sleep time. There is no consensus among the experts on which of these strategies is best. Most people learn through trial-and-error what works for them. Social factors play a big role in determining the time for sleep. Some people may have to negotiate sleep time with their families. A worker who has to spend most of his or her days off catching up on sleep is not getting enough sleep the rest of the time.

- ***Use anchor sleep to ease schedule changes.*** Sleep as much as possible at the same time on both the shift you are coming off and the one you are going on. For example, at least two days before rotating from days to evenings, and from sleeping from midnight to 6 A.M., start sleeping from 3 to 9 A.M. The period from 3 to 6 A.M. serves as anchor sleep. It will help keep you on an even keel. Try to carry a two- to four-hour period of anchor sleep with you at each transition, and keep it on days off as well.

  Before rotating to a night shift, go to bed as late as you can at night and sleep as late as possible in the mornings. In the morning, wear sunglasses outside or stay indoors.

  When coming off the night shift, returning to being a day person should be easy. After the last night shift, sleep as little as you can tolerate, and then stay awake until a conventional bedtime. You should fall asleep quickly that night.

  When on the night shift, if you rotate in a forward direction, wear sunglasses on the way home from work. If you rotate backward, spend fifteen minutes or more outdoors after work.

- ***Protect your sleep.*** If you sleep during the day, darken the bedroom windows with opaque shades or curtains, black vinyl, aluminum foil, or Styrofoam cutouts. Find a quiet place to sleep. Some shift workers set up a bedroom in the basement. To minimize noise, unplug the phone or use an answering machine. Ask friends not to call during sleep hours. Try earplugs. Consider installing thick carpets in your bedroom and sound-proofing tape around the bedroom door.

  Put a sign on the doorbell saying DO NOT RING; DAY SLEEPER, or install a switch that lets you turn off the doorbell.

- ***Pay attention to what, and when, you eat.*** Regardless of what the clock says, eat the appropriate foods at breakfast time on your work schedule, and so on. Try to eat breakfast soon after you get out of bed. As far as your biologic clock is concerned, breakfast is the most important meal of the day: it tells your body that the day is starting.

  Have lunch midway through your shift. Do something active on your lunch break: take a walk or play a game of darts.

  When working evenings or nights, skip chili, pepperoni pizza, and other hard-to-digest foods at dinner. Eat lightly before bedtime on any shift.

  Try to join your family for at least one meal a day, even if it is your supper and their breakfast. Some foods, like omelets and potatoes, feel right at any time of day.

  On your days off, shift mealtimes as you do sleep times.

- ***Use caffeine judiciously.*** Avoid caffeine within four or five hours of bedtime on all shifts.

- ***Do not drink alcohol to induce sleep.*** For people who must sleep during the day, staying asleep is the problem, not getting to sleep. Alcohol makes sleep shorter and more restless.

- ***Do not smoke at bedtime.*** Nicotine is a stimulant.

- ***Limit use of sleeping pills.*** They may help you sleep a little longer for the first couple of sleep periods after a shift change, but they are not designed for daily use.

- ***Take extra care driving home.*** Be especially careful if it is dark. Traffic accidents occur three times more often in the dark.

- ***Develop personalized zeitgebers.*** Tell your body "It's morning," or "It's night," with personal rituals. Lay out the next day's clothes on a chair, put your keys by the door, assign a certain time of day to read the newspaper.

- ***Post a calendar in the kitchen at home.*** Use different colors of ink, or colored stickers, to show which days you will be on different schedules.

  Schedule family time together. Make a date with your spouse for dinner, a movie, or simply taking a walk together, and write it on the calendar. Get your children's after-school sports schedule. Plan to attend a particular game and mark it on the calendar for everyone to look forward to.

  Make the effort to attend family events even if you have to arrive late or leave early. Ask someone to take photos or videotape events that you cannot attend.

- ***Keep family communication channels open.*** Buy a bulletin board or magnets for the refrigerator, so family members who do not see each other on workdays can leave notes. Do not just leave lists of chores for each other or for the kids. "I heard dance practice went great," or "Nice grade on that spelling test," goes a long way toward making a parent's presence felt, even if that parent is not available for a few days.

- ***Improve home security.*** To minimize worries about the family's safety at night, consider installing a security system or getting a dog. Leaving some lights on may also alleviate some concerns.

- ***Exercise regularly.*** Jogging, aerobics, and other vigorous exercise soon after you awaken raises body temperature and makes you feel more alert sooner. Like setting regular times for sleep and meals, exercise helps synchronize body clocks.

  Even if your job requires physical labor, a regular exercise program can boost your stamina. Most people enjoy these activities more with company. Avoid doing exercise within an hour or two of bedtime. It keeps body temperature high and makes falling asleep more difficult.

- ***Talk with your doctor if you have a chronic illness.*** Irregular hours aggravate some illnesses. That does not mean you have to avoid shift work, but you will probably need to take special precautions. If you have diabetes, for example, you must schedule meals and snacks carefully. If you have epilepsy, you need to get enough sleep to avert seizures. If you have narcolepsy, you need to avoid many monotonous or repetitive tasks that may exacerbate sleepiness as well as those in which sleepiness may put you or others in danger.

HELP YOUR BABY SLEEP THROUGH THE NIGHT

- ***Do not worry about schedules in the first months of your baby's life.*** The baby needs the security of knowing you are available at any time, and will learn quickly that night is the time for sustained sleep. After about three months, you can develop a standard bedtime ritual for the baby.

- ***Schedule a quiet time for an hour or so before sleep.*** Don't shake rattles, bounce the baby on your knee, or engage in other spirited play.

- ***Put the baby into bed.*** Sing if you like, but make it short.

- ***Turn off the lights (except for a night-light) and leave the room.***

- ***If the baby cries, go in but do not turn on the lights.*** Pat the baby, but avoid taking the baby out of bed. Keep talking to a minimum. Your aim is to teach your baby that it is time to sleep, and that the bed is for sleeping.

HOW TO COPE WITH PMS

Although dozens of treatments have been proposed for PMS, none has proven universally effective. Many women say these measures help:

- ***Assess your own rhythms and factor them into daily life.*** If you need extra support and understanding at a particular time of month from a partner or even co-workers, ask for it. If you have a choice, avoid scheduling important business meetings or other major events at this time. Stress does not cause PMS, but it may make the symptoms worse.

- ***Restrict caffeine intake.*** Excess amounts of caffeine may trigger irritability, insomnia, and an upset stomach at any time, so it would be wise to limit your intake when these symptoms are particularly likely to be a problem. Coffee, tea, cola drinks, many over-the-counter drugs (including pain relievers and cold remedies), and chocolate, all contain caffeine or related compounds.

- ***Consume complex carbohydrates.*** Whole grains, breads, pasta, potatoes, and breakfast cereals may boost serotonin. High amounts of this brain chemical are thought to have a calming effect, while low amounts are linked with depression and anxiety. In a study at the Massachusetts Institute of Technology, women with PMS said that their mood improved after they ate a large bowl of cornflakes.

- ***Minimize salt intake.*** Salt may cause you to retain fluids, leading to weight gain, bloating, ankle swelling, headaches, and other PMS symptoms. Most diet sodas, fast-food meals, packaged luncheon meats, and other convenience foods are high in sodium. If you gain more than two or three pounds before your periods, ask your doctor if taking a diuretic could help.

- ***Restrict alcohol intake.*** Alcohol may foster feelings of depression and hopelessness. Some women report a decreased tolerance for alcohol premenstrually.

- ***Exercise regularly.*** While the benefits of exercise to alleviate PMS symptoms remain unproven, exercise does contribute to overall health. It may improve mood, too, by triggering the release of endorphins, the body's natural painrelievers.

- ***Get enough sleep.*** Some women report excessive sleepiness right before their periods. Lack of sleep itself may trigger irritability.

- ***If your symptoms interfere with work or social life, see your doctor.*** Numerous drugs may provide relief. The choice will depend on your specific symptoms.

SAMPLE SLEEP/WAKE DIARY

Use the three-day sample below as a guide to prepare your sleep/wake diary. You will need to keep a diary for at least a week to get a good sense of your sleep patterns. Record all times in hours and minutes, e.g., 6:15. Mark all sleep periods, including naps, with a dark line. Note events or activities that seem to harm or help your sleep—coffee, alcohol, stressful day, day off from work, argument with spouse, trip, and, for women, time of menstrual cycle.

Sleep Period Graph. Mark with a horizontal line all the times that you are asleep, including naps. Use an arrow pointing down to show when you get into bed. Mark the times you get out of bed with an arrow pointing up.

Use the following abbreviations:

M: meals
S: snacks
P: sleeping pills
D: other drugs (number and list separately)
X: exercise
A: alarm clock

Day of the Week. Write in the day of the week for the *night* on which that 24-hour period began. For example, if the first 24-hour period is Tuesday night to Wednesday evening, write "Tuesday" in the first blank.

Clock time: into bed. Write in the time that you first got into bed.

Clock time: attempted to sleep. Write in the time that you first began trying to sleep.

Minutes to fall asleep. This asks you to estimate how long it took you to fall asleep from the time you first began trying to sleep.

Total hours slept. Estimate the total amount of sleep you received, excluding wake time during the night and naptime during the day.

Name Niki Ryan

Week beginning Sept. 5, 1994

↓ = in bed ↑ = out of bed

SAMPLE SLEEP/WAKE DIARY, *Continued*

PM Midnight AM Noon PM
6 7 8 9 10 11 12 1 2 3 4 5 6 7 8 9 10 11 12 1 2 3 4 5

Sleep period graph M ↓ ↑ M M X
WATCH TV IN BED A

Day of the week: MON. night
Clock time: Into bed 11 PM
Clock time: Attempted to sleep 11 30 PM
Minutes to fall asleep 60
Total hours slept 6
On a scale of 0 (very poor) to 100 (very good), my sleep was 55
On a scale of 0 (exhausted) to 100 (refreshed), this morning I felt 70

PM Midnight AM Noon PM
6 7 8 9 10 11 12 1 2 3 4 5 6 7 8 9 10 11 12 1 2 3 4 5

Sleep period graph M P ↓ ↑ ↓ ↑ M M ↓ ↑
Read book A Nap

Day of the week: Tues. night
Clock time: Into bed 11 15
Clock time: Attempted to sleep 11 15
Minutes to fall asleep 15
Total hours slept 4.5
On a scale of 0 (very poor) to 100 (very good), my sleep was 40
On a scale of 0 (exhausted) to 100 (refreshed), this morning I felt 30

PM Midnight AM Noon PM
6 7 8 9 10 11 12 1 2 3 4 5 6 7 8 9 10 11 12 1 2 3 4 5

Sleep period graph M ↓ ↑ ↓ ↑ S ↓ ↑
Alcohol Nap - had the Flu

Day of the week: Wed. night
Clock time: Into bed 12 30 AM
Clock time: Attempted to sleep 12 30 AM
Minutes to fall asleep 30
Total hours slept 7½
On a scale of 0 (very poor) to 100 (very good), my sleep was 65
On a scale of 0 (exhausted) to 100 (refreshed), this morning I felt 25

Adapted from the Stanford University Sleep Diary Log

PERSONAL SLEEP/WAKE DIARY

Name ________________ M = normal meal P = sleeping pill A = alarm
Week S = snack D = other drug X = exercise
beginning ________, 19 __ ↓ = in bed ↑ = out of bed

PM Midnight AM Noon PM
6 7 8 9 10 11 12 1 2 3 4 5 6 7 8 9 10 11 12 1 2 3 4 5
Sleep period graph

Day of the week: ________ night
Clock time: Into bed ________
Clock time: Attempted to sleep ________
Minutes to fall asleep ________
Total hours slept ________
On a scale of 0 (very poor) to 100 (very good), my sleep was ________
On a scale of 0 (exhausted) to 100 (refreshed), this morning I felt ________

PM Midnight AM Noon PM
6 7 8 9 10 11 12 1 2 3 4 5 6 7 8 9 10 11 12 1 2 3 4 5
Sleep period graph

Day of the week: ________ night
Clock time: Into bed ________
Clock time: Attempted to sleep ________
Minutes to fall asleep ________
Total hours slept ________
On a scale of 0 (very poor) to 100 (very good), my sleep was ________
On a scale of 0 (exhausted) to 100 (refreshed), this morning I felt ________

PM Midnight AM Noon PM
6 7 8 9 10 11 12 1 2 3 4 5 6 7 8 9 10 11 12 1 2 3 4 5
Sleep period graph

Day of the week: ________ night
Clock time: Into bed ________
Clock time: Attempted to sleep ________
Minutes to fall asleep ________
Total hours slept ________
On a scale of 0 (very poor) to 100 (very good), my sleep was ________
On a scale of 0 (exhausted) to 100 (refreshed), this morning I felt ________

PM Midnight AM Noon PM
6 7 8 9 10 11 12 1 2 3 4 5 6 7 8 9 10 11 12 1 2 3 4 5
Sleep period graph

PERSONAL SLEEP/WAKE DIARY, *Continued*

Day of the week: ________ night

Clock time: Into bed ________
Clock time: Attempted to sleep ________

Minutes to fall asleep ________
Total hours slept ________

On a scale of 0 (very poor) to 100 (very good), my sleep was ________
On a scale of 0 (exhausted) to 100 (refreshed), this morning I felt ________

PM Midnight AM Noon PM

	6	7	8	9	10	11	12	1	2	3	4	5	6	7	8	9	10	11	12	1	2	3	4	5
Sleep period graph																								

Day of the week: ________ night

Clock time: Into bed ________
Clock time: Attempted to sleep ________

Minutes to fall asleep ________
Total hours slept ________

On a scale of 0 (very poor) to 100 (very good), my sleep was ________
On a scale of 0 (exhausted) to 100 (refreshed), this morning I felt ________

PM Midnight AM Noon PM

	6	7	8	9	10	11	12	1	2	3	4	5	6	7	8	9	10	11	12	1	2	3	4	5
Sleep period graph																								

Day of the week: ________ night

Clock time: Into bed ________
Clock time: Attempted to sleep ________

Minutes to fall asleep ________
Total hours slept ________

On a scale of 0 (very poor) to 100 (very good), my sleep was ________
On a scale of 0 (exhausted) to 100 (refreshed), this morning I felt ________

PM Midnight AM Noon PM

	6	7	8	9	10	11	12	1	2	3	4	5	6	7	8	9	10	11	12	1	2	3	4	5
Sleep period graph																								

Day of the week: ________ night

Clock time: Into bed ________
Clock time: Attempted to sleep ________

Minutes to fall asleep ________
Total hours slept ________

On a scale of 0 (very poor) to 100 (very good), my sleep was ________
On a scale of 0 (exhausted) to 100 (refreshed), this morning I felt ________

SLEEPINESS/ALERTNESS CHART

INSTRUCTIONS:

1. Complete this chart during the time that you are keeping the diary.
2. Select two typical CONSECUTIVE working days.
3. Write in the days and dates you selected at the top of each chart.
4. Throughout your waking hours on these days, RATE yourself every two hours according to the scale at the left of the chart.
5. CIRCLE the number indicating your level of alertness or sleepiness for each two-hour time period.
6. CIRCLE "X" for those times when you were asleep.

DAY ONE — Day ______ Date ______

DAY TWO — Day ______ Date ______

	MID-	AM PERIOD						PM PERIOD					MID-	AM PERIOD						PM PERIOD					MID-
SCALE	NIGHT	2	4	6	8	10	NOON	2	4	6	8	10	NIGHT	2	4	6	8	10	NOON	2	4	6	8	10	NIGHT
1. Alert. Wide awake. Energetic.	1	1	1	1	1	1	1	1	1	1	1	1	1	1	1	1	1	1	1	1	1	1	1	1	1
2. Functioning at a high level, but not at a peak. Able to concentrate.	2	2	2	2	2	2	2	2	2	2	2	2	2	2	2	2	2	2	2	2	2	2	2	2	2

3. Awake, but not fully alert.	3	3 3 3 3 3	3	3 3 3 3 3	3	3 3 3 3 3	3	3 3 3 3 3	3
4. A little foggy, let down.	4	4 4 4 4 4	4	4 4 4 4 4	4	4 4 4 4 4	4	4 4 4 4 4	4
5. Foggy. Beginning to lose interest in remaining awake. Slowed down.	5	5 5 5 5 5	5	5 5 5 5 5	5	5 5 5 5 5	5	5 5 5 5 5	5
6. Sleepy. Prefer to be lying down. Woozy.	6	6 6 6 6 6	6	6 6 6 6 6	6	6 6 6 6 6	6	6 6 6 6 6	6
7. Cannot stay awake. Sleep onset soon.	7	7 7 7 7 7	7	7 7 7 7 7	7	7 7 7 7 7	7	7 7 7 7 7	7
X. Asleep.	X	X X X X X	X	X X X X X	X	X X X X X	X	X X X X X	X

Adapted from the Stanford University Sleepiness Scale

GLOSSARY*

Actigraph. A biomedical instrument for the measurement of body movement.

Arise time. The clock time that a person gets out of bed after sleep. Distinguished from final wake-up.

Arousal. An abrupt change from a deeper stage of NREM sleep to a lighter stage, or from REM sleep toward wakefulness, with the possibility of awakening as the final outcome. Arousal may be accompanied by increased muscle activity and heart rate, as well as body movements.

Awakening. The return to the polysomnographically defined awake state from any sleep stage, paralleled by a resumption of a reasonably alert state of environmental awareness.

Bedtime. The clock time when a person attempts to fall asleep, as differentiated from the clock time of getting into bed.

Chronobiology. The science relating to temporal, primarily rhythmical, processes in biology.

Circadian rhythm. An innate, daily fluctuation of physiological or behavioral functions, including sleep/wake states. Generally tied to the twenty-four-hour daily dark/light cycle. Sometimes occurs at a measurably different (e.g., twenty-three or twenty-five-hour) periodicity when light/dark and other time cues are removed.

Cycle. Characteristic of an event exhibiting rhythmic fluctuations. One cycle is defined as the activity from one maximum, or minimum, to the next.

Delayed sleep phase. A condition that occurs when the clock hour at which sleep normally occurs is moved back in time within a given twenty-four-hour sleep/wake cycle. This results in temporarily displaced, or delayed,

* Adapted, with permission, from the *International Classification of Sleep Disorders: Diagnostic and Coding Manual.* Diagnostic Classification Steering Committee: Michael Thorpy, Chairman. Rochester, Minnesota: American Sleep Disorders Association, 1990.

sleep within the twenty-four-hour cycle. The same term denotes a circadian rhythm sleep disturbance called the *delayed sleep phase syndrome*.

Diurnal. Pertaining to the daytime.

Early-morning arousal. A premature morning awakening.

Electroencephalogram (EEG). A recording of the electrical activity of the brain by means of electrodes placed on the surface of the head, one of the basic variables used to score sleep stages and waking.

Entrainment. Synchronization of a biologic rhythm by a forcing stimulus such as an environmental time cue, or zeitgeber.

Excessive sleepiness (-somnolence, hypersomnia, excessive daytime sleepiness). A subjective report of difficulty in maintaining the alert awake state, usually accompanied by a rapid entrance into sleep when the person is sedentary. May be due to an excessively deep or prolonged major sleep episode. Can be quantitatively measured by use of subjectively defined rating scales of sleepiness or by electrophysiological tests, such as **MSLT**. (See **Multiple Sleep Latency Test.**). Most commonly occurs during the daytime, but may be present at night in a person who has his or her major sleep episode during the daytime, such as a shift worker.

Final awakening. The amount of wakefulness that occurs after the final wake-up time until the arise time (lights on).

Final wake-up. The clock time at which an individual awakens for the last time before the arise time.

Fragmentation. The interruption of any stage of sleep due to the appearance of another sleep stage, or to wakefulness, leading to disrupted NREM-REM sleep cycles. Sleep fragmentation connotes repetitive interruptions of sleep by arousals and awakenings.

Free-running. A chronobiologic term that refers to the natural endogenous period of a rhythm when zeitgebers are removed. In humans, it most commonly appears as the tendency to delay some circadian rhythms, such as the sleep/wake cycle, by approximately one hour every day, when a person has an impaired ability to entrain, or is without time cues.

Hypnagogic. Refers to the occurrence of an event during the transition from wakefulness to sleep.

Hypnagogic imagery (hallucinations). Vivid sensory images occurring at sleep onset but particularly vivid with sleep-onset REM periods. A feature of narcoleptic naps, when sleep starts with REM sleep.

Hypnagogic startle. A "sleep start," or sudden body jerk (hypnic jerk), observed normally just at sleep onset and usually resulting, at least momentarily, in an awakening.

Insomnia. Difficulty in initiating and/or maintaining sleep. Term is employed ubiquitously to indicate any and all gradations and types of sleep loss.

Into bed time. The clock time at which a person gets into bed. The into bed time (IBT) will be the same as actual bedtime for many people, but not for those who spend time in wakeful activities, such as reading in bed, before attempting to sleep.

Light/dark cycle. The periodic pattern of light, artificial or natural, alternating with darkness.

Maintenance of Wakefulness Test (MWT). A series of measurements of the interval from lights out to sleep onset, utilized in the assessment of the ability to remain awake. Subjects are instructed to try to remain awake in a darkened room while in a semireclining position. This test is most useful for assessing the effects of medication upon the ability to remain awake.

Major sleep episode. The longest sleep episode that occurs on a daily basis. Typically dictated by the circadian rhythm of sleep and wakefulness, it is the conventional or habitual time for sleeping.

Microsleep. An episode lasting up to thirty seconds during which external stimuli are not perceived. Microsleeps are associated with excessive sleepiness and automatic behavior.

Multiple Sleep Latency Test (MSLT). A series of measurements of the interval from lights out to sleep onset. Subjects are allowed a fixed number of opportunities to fall asleep during their customary awake period. Excessive sleepiness is characterized by short latencies. Long latencies help distinguish physical tiredness or fatigue from true sleepiness.

Nap. A short sleep episode intentionally or unintentionally taken during the period of habitual wakefulness.

Nocturnal penile tumescence (NPT). The natural periodic cycle of penile erections that occur during sleep, typically associated with REM sleep. Preferred term: sleep-related erections.

Nocturnal sleep. The typical nighttime or major sleep episode related to the circadian rhythm of sleep and wakefulness; also the conventional or habitual time for sleeping.

NREM-REM sleep cycle (synonymous with sleep cycle). An episode during sleep composed of a NREM sleep episode and the subsequent REM sleep episode. Each NREM-REM sleep couplet is equal to one cycle. Any NREM sleep stage suffices as the NREM sleep portion of a cycle. An adult sleep period of six and one half to eight and one half hours generally consists of four to six cycles. The cycle duration increases from infancy to young adulthood.

NREM sleep period. The NREM sleep portion of the NREM-REM sleep cycle. Such an episode consists primarily of Sleep Stages 3-4 early in the night and of Sleep Stage 2 later on. (See **Sleep Stages.**)

Period. The time that elapses before a rhythm starts to repeat itself—that is, between one peak or trough and the next.

Phase advance. A shift of an episode of sleep or wake to an earlier position in the twenty-four-hour sleep/wake cycle. A shift in sleep from 11 P.M.–7 A.M. to 8 P.M.–4 A.M. represents a three-hour phase advance.

Phase delay. A shift of an episode of sleep or wake to a later position in the twenty-four-hour sleep/wake cycle. It is the exact opposite of phase advance. These terms differ from common concepts of change in clock time: to effect a phase delay, the clock advances. To effect a phase advance, the clock moves backward.

Phase transition. One of the two junctures of the major sleep and wake phases in the twenty-four-hour sleep/wake cycle.

Photoperiod. The duration of light in a light/dark cycle.

Polysomnogram. The continuous and simultaneous recording of brain waves, eye movements, and muscle activity during sleep.

Premature morning awakening. Early termination of the sleep episode with inability to return to sleep. It reflects interference at the end rather than at the beginning of the sleep episode. It is a characteristic sleep disturbance in many people with depression.

Rhythm. An event that occurs at an approximately constant period length.

Sleep efficiency (or sleep efficiency index). The ratio of total sleep time to time in bed.

Sleep episode. An interval of sleep that may be voluntary or involuntary. In the sleep laboratory, the sleep episode occurs from the time of lights out to the time of lights on. The major sleep episode is usually the longest daily sleep episode.

Sleep hygiene. Conditions and practices that promote continuous and effective sleep. These include regularity of bedtime and arise time, conformity of time spent in bed to the time necessary to avoid sleepiness when awake, restriction of alcohol and caffeine beverages prior to bedtime, and employment of exercise, nutrition, and environmental factors to enhance restful sleep.

Sleepiness (or somnolence, drowsiness). Difficulty in maintaining alert wakefulness. If not kept actively aroused, a person falls asleep. This is not simply a feeling of physical tiredness or listlessness. Sleepiness in inappropriate circumstances (e.g., driving) is excessive sleepiness.

Sleep interruption. Breaks in sleep resulting in arousal and wakefulness.

Sleep latency. The duration of time from lights out, or bedtime, to the start of sleep.

Sleep log (diary). A daily written record of a person's sleep/wake pattern, containing information such as time of retiring and arising, time in bed, estimated total sleep time, number and duration of sleep interruptions, quality of sleep, daytime naps, use of medications or caffeine, and nature of waking activities.

Sleep mentation. The imagery and thinking experienced during sleep, usually during REM sleep. Imagery is vividly expressed in dreams involving all the senses in approximate proportion to their waking representations. Mentation is experienced generally less distinctly in NREM sleep, but it may be quite vivid in Stage 2 sleep, especially toward the end of the episode. Mentation at sleep onset (hypnagogic reverie) may be as vivid as in REM sleep.

Sleep onset. The transition from awake state to sleep state, normally to NREM Stage 1 sleep, but in certain conditions, such as during infancy and in people with narcolepsy, into REM sleep.

Sleep pattern (twenty-four-hour sleep/wake pattern). A person's clock-hour schedule of bedtime and arise time as well as nap behavior. May also include time and duration of sleep interruptions.

Sleep Stage REM. Named for distinctive spontaneous rapid eye movements. The stage of sleep with the highest brain activity, characterized by enhanced brain metabolism and vivid hallucinatory imagery, or dreaming. Awakening threshold to nonsignificant stimuli is high. REM sleep is usually 20 to 25 percent of total sleep time. Because resting muscle activity is suppressed, it also is called *paradoxical sleep*.

Sleep Stages. Distinctive stages of sleep, best demonstrated by polysomnographic recordings.

Sleep Stage 1 (NREM Stage 1). Occurs at sleep onset or follows arousal from Sleep Stages 2, 3-4, or REM. Stage 1 normally represents 4 to 5 percent of the major sleep episode.

Sleep Stage 2 (NREM Stage 2). Light sleep. Stage 2 usually accounts for 45 to 55 percent of the major sleep episode.

Sleep Stage 3 (NREM Stage 3). Deeper sleep. With Stage 4, it constitutes deep NREM sleep, the so-called "slow-wave sleep." Stage 3 usually appears only in the first third of the sleep episode and comprises 4 to 6 percent of total sleep time. It is often combined with Stage 4 into NREM Sleep Stage 3-4 because of the lack of documented physiological differences between the two.

Sleep Stage 4 (NREM Stage 4). Deepest sleep. Stage 4 usually represents 12 to 15 percent of total sleep time. Sleepwalking, sleep terrors, and confusional arousal episodes generally start in Stage 4 or during arousals from this stage.

Sleep/wake cycle. The clock-hour relationships of the major sleep/wake episodes in the twenty-four-hour cycle.

Sleep/wake shift (change, reversal). When sleep as a whole, or in part, moves to a time of customary waking activity, and the latter moves to the time of the major sleep episode. Common in jet lag and shift work.

Synchronized. When two or more rhythms recur with the same phase relationship.

Total sleep episode. The total time available for sleep during an attempt to sleep. It comprises NREM and REM sleep as well as wakefulness.

Total sleep time (TST). The amount of actual time spent asleep in a sleep episode.

Wake time. The total time scored as wakefulness in a polysomnogram between sleep onset and final wake-up.

Zeitgeber. An environmental time cue that usually helps entrainment to the twenty-four-hour day. Zeitgebers include sunlight, noise, social interaction, and alarm clocks.

REFERENCES

Chapter 1: The Biologic Clockshop

Adler, Stephen, and Stephen Wermiel. India's Justices Uphold $465 Million Bhopal Settlement. *The Wall Street Journal*, January 2, 1990:A2.

Aschoff, Jürgen. Annual Rhythms in Man. In *Handbook of Behavioral Neurobiology, Volume 4: Biological Rhythms*, ed, Jürgen Aschoff. New York: Plenum Press, 1981, pp. 475–487.

Biological Clocks and Shift Work Scheduling. Hearings before the Subcommittee on Investigations and Oversight of the Committee on Science and Technology, House of Representatives, Ninety-eighth Congress, March 23–24, 1983. Washington, D.C.: U.S. Government Printing Office, 1983, p. 171.

Bunning, Erwin. Opening Address: Biological Clocks. Cold Spring Harbor Symposium. *Quantitative Biology*, 1960;25:1–9D.

Carroll, Lewis. *The Adventures of Alice in Wonderland* and *Through the Looking-Glass.* New York: New American Library, 1960, p. 47.

Chaykin, S. Beard Growth: A Window for Observing Circadian and Infradian Rhythms of Men. *Chronobiologia*, 1986;13:163–165.

de Mairan, Jean-Jacques d'Ortous. Observation botanique. *Histoire de l'Academie Royale des Sciences*, Paris, 1729, p. 35.

Ehlers, Cindy, et al. Social Zeitgebers and Biological Rhythms. *Archives of General Psychiatry*, 1988;45:948–952.

Feynman, Richard. *What Do You Care What Other People Think?* New York: Norton, 1988.

Guilleminault, Christian, et al. Development of Temperature Rhythm in Normal Infants. *Sleep Research*, 1987;17:611.

Halberg, Franz. Physiologic 24-Hour Periodicity; General and Procedural Considerations with Reference to the Adrenal Cycle. *Vitamin, Hormon un Fermentforschung.* 1959;3/4:225–297.

IAEA Conference Studies: Chernobyl A-Plant Accident. *Facts on File.* New York: Facts on File, Inc., August 29, 1986, p. 634.

Kittrell, Melanie. Interview, October 1988.

Levine, Richard, et al. Differences in the Quality of Semen in Outdoor Workers During Summer and Winter. *New England Journal of Medicine*, 1990;323:12–16.

Lewy, Alfred, et al. Bright Artificial Light Treatment of a Manic-Depressive Patient with a Seasonal Mood Cycle. *American Journal of Psychiatry*, 1982;139:1496–97.

Marine Accident Report: Grounding of the U.S. Tankship Exxon Valdez on Bligh Reef, Prince William Sound near Valdez, Alaska, March 24, 1989. Washington, D.C.: National Transportation Safety Board, 1990, pp. 128, 166.

Mills, J. N. Circadian Rhythms During and After Three Months in Solitude Underground. *Journal of Physiology*, 1964;174:217–231.

Moore, R. Y., and V. B. Eichler. Loss of Circadian Adrenal-Corticosterone Rhythm Following Suprachiasmatic Lesions in the Rat. *Brain Research*, 1972;42:201–206.

Moore-Ede, Martin, Frank Sulzman, and Charles Fuller. *The Clocks That Time Us: Physiology of the Circadian Timing System.* Cambridge: Harvard University Press, 1982, p. 61.

———. *The Twenty-Four Hour Society: Understanding Human Limits in a World That Never Stops.* Reading, Mass.: Addison-Wesley, 1993, p. 109.

National Commission on Sleep Disorders Research. *Wake Up America: A National Sleep Alert.* Volume 1: Executive Summary and Executive Report. Bethesda, Md.: National Institutes of Health, 1993, p. 43.

Ralph, Martin, et al. Transplanted Suprachiasmatic Nucleus Determines Circadian Period. *Science*, 1990;247:975–978.

Report of the Presidential Commission on the Space Shuttle Challenger Accident. Vol. II, Appendix G—Human Factors Analysis. Washington, D.C.: U.S. Government Printing Office, 1986, pp. G–1, G–5.

Report of the President's Commission on the Accident at Three Mile Island. Washington, D.C.: U.S. Government Printing Office, 1979, p. 8.

Richter, Curt P. A Behavioristic Study of the Activity of the Rat. Comparative Psychology Monographs, Vol. 1, Serial 2, September 1922. In *The Psychobiology of Curt Richter*, ed., Elliot Blass. Baltimore: York Press, 1976, pp. 3–49.

———. *Biological Clocks.* Springfield, Ill.: Charles C. Thomas, Publisher, 1965, p. 96.

———. Interview, January 2, 1987.

———. It's a Long Way to Tipperary, the Land of My Genes. In *Leaders in the Study of Animal Behavior: Autobiographical Perspectives*, ed., D. A. Dewsbury. Lewisburg, Pa.: Bucknell University Press, 1985, pp. 356–386.

Rozin, Paul. Curt Richter: The Compleat Psychobiologist. In *The Psychobiology of Curt Richter*, ed., Elliot Blass. Baltimore: York Press, 1976, pp. xv–xxviii.

Smolensky, Michael. Aspects of Human Chronopathology. In *Biological Rhythms and Medicine*, eds., Alain Reinberg and Michael Smolensky. New York: Springer-Verlag, 1983, pp. 131–209.

Stephan, F. K., and I. Zucker. Circadian Rhythms in Drinking Behavior and Locomotor Activity of Rats Are Eliminated by Hypothalamic Lesions. *Proceedings of the National Academy of Sciences (USA)*, 1972;69:1583–1586.

United States Coast Guard. *Report of the Tanker Safety Study Group.* Washington, D.C.: U.S. Department of Transportation, October 6, 1989, p. 32.

U.S. Congress, Office of Technology Assessment. *Biological Rhythms: Implications for the Worker.* OTA-BA-463. Washington, D.C.: U.S. Government Printing Office, September 1991, pp. iii, 199.

Vital and Health Statistics. Seasonal Variation of Births, United States 1933–63. Washington, D.C.: U.S. Department of Health, Education and Welfare, Series 21, No. 9, 1966.

Vital and Health Statistics. Characteristics of Births 1973–75; Washington, D.C.: U.S. Department of Health and Human Services, Series 21, No. 30, 1978.

Webb, Wilse. Foreword. In *Rhythmic Aspects of Behavior*, eds., Frederick M. Brown and R. Curtis Graeber. Hillsdale, N.J.: Lawrence Erlbaum Associates, 1982, pp. ix–xiii.

Wever, Rütger. Light Effects on Human Circadian Rhythms: A Review of Recent Andechs Experiments. *Journal of Biological Rhythms,* 1989; 4: 161–185.

Winstead, Daniel, et al. Biorhythms: Fact or Superstition? *American Journal of Psychiatry*, 1981;138(9):1188–1192.

Zerubavel, Eviatar. *The Seven Day Circle.* New York: The Free Press, 1985.

Chapter 2. Life Without Time Cues

Kleitman, Nathaniel. *Sleep and Wakefulness.* Chicago: University of Chicago Press, 1963, pp. 178–182.

Tempest, Rone. The Longest Night. *Los Angeles Times*, December 9, 1988, Part V.

Weitzman, Elliot, Charles Czeisler, and Martin Moore-Ede. Sleep-Wake, Neuroendocrine and Body Temperature Circadian Rhythms Under Entrained and Non-Entrained (Free-Running) Conditions in Man. In *Biological Rhythms and Their Central Mechanism*, eds., Masami Suda, Osamu Hayaishi, and Hachiro Nakagawa. Elsevier/North-Holland Biomedical Press, 1979, pp. 199–227.

Chapter 3: Day and Night: The Rhythms of Alertness

Adams, B. N. Interview, April 23, 1989.

———. and R. Cromwell. Morning and Night People in the Family: A Preliminary Statement. *Family Relations*, 1978;27:5–13.

Craig, Angus. Acute Effects of Meals on Perceptual and Cognitive Efficiency. Presented at a conference: *Diet and Behavior: A Multidisciplinary Evaluation*, sponsored by the American Medical Association, International Life Sciences Institute, and the Nutrition Foundation, Inc., November 27–29, 1984.

Folkard, Simon. Time of Day Effects in Schoolchildren's Immediate and Delayed Recall of Meaningful Material. *British Journal of Psychology*, 1977;68:45–50.

Giambra, Leonard, et al. A Circadian Rhythm in the Frequency of Spontaneous Task-Unrelated Images and Thoughts. *Imagination, Cognition, and Personality*, 1988–1989;8:309–314.

Goleman, Daniel. Embattled Giant of Psychology Speaks His Mind. *The New York Times*, August 25, 1987: C1, C3.

Kleitman, Nathaniel. Basic Rest-Activity Cycle—22 Years Later. *Sleep*, 1982;5:311–317.

Kripke, Daniel. Interview, May 8, 1989.

———. Ultradian Rhythms in Behavior and Physiology. In *Rhythmic Aspects of Behavior*, eds., Frederick M. Brown and R. Curtis Graeber. Hillsdale, N.J.: Lawrence Erlbaum Associates, 1982, pp. 313–344.

Lamberg, Lynne. It's About Time. *American Way*, September 16, 1986:82–87.

Larson, Jeffry, et al. Morning and Night Couples: The Effect of Wake and Sleep Patterns on Marital Adjustment. *Journal of Marital and Family Therapy*, 1991;17:53–65.

Lieberman, Harris. Quantifying the Behavioral Effects of Food Constituents. Presented at a conference: *Diet and Behavior: A Multidisciplinary Evaluation*, sponsored by the American Medical Association, International Life Sciences Institute, and the Nutrition Foundation, Inc., November 27–29, 1984.

Melbin, Murray. *Night as Frontier: Colonizing the World After Dark.* New York: The Free Press, 1987.

Mitler, Merrill, et al. Catastrophes, Sleep, and Public Policy: Consensus Report. *Sleep*, 1988;11:100–109.

Monk, Timothy, and Simon Folkard. Concealed Inefficiency of Late-Night Study. *Nature*, 1978;273(5660):296–297.

Monk, Timothy, et al. Circadian Factors During Sustained Performance: Background and Methodology. *Behavior Research Methods, Instruments and Computers*, 1985;17(1):19–26.

National Commission on Sleep Disorders Research. *Wake-Up America: A N-*

tional Sleep Alert. Vol. 1: Executive Summary and Executive Report. Bethesda, Md.: National Institutes of Health, 1993, p. vi.

Chapter 4: Night and Day: The Rhythms of Sleep

Allen, Richard. Social Factors Associated with the Amount of School Week Sleep Lag for Seniors in an Early Starting Suburban High School. *Sleep Research*, 1992;21:114.

Allen, Richard, and Jerome Mirabile. Self-Reported Sleep-Wake Patterns During the School Year from Two Different Senior High Schools. *Sleep Research*, 1989;18:132.

Anonymous. It's Over, Debbie. *Journal of the American Medical Association*, 1988;259:272.

Aserinsky, E., and N. Kleitman. Regularly Occurring Periods of Eye Motility, and Concomitant Phenomena During Sleep. *Science*, 1953;118:273–274.

Biddle, Jeff, and Daniel Hamermesh. *Sleep and the Allocation of Time. Working Paper No. 2988.* Cambridge, Mass.: National Bureau of Economic Research, 1989.

Broughton, Roger. Chronobiological Aspects and Models of Sleep and Napping. In *Sleep and Alertness: Chronobiological, Behavioral, and Medical Aspects of Napping*, eds., David Dinges and Roger Broughton. New York: Raven Press, 1989, pp. 71–98.

———. Interview, June 1986.

———. Human Consciousness and Sleep/Waking Rhythms: A Review and Some Neuropsychological Considerations. *Journal of Clinical Neuropsychology*, 1982;4:193–218.

Carskadon, Mary. Patterns of Sleep and Sleepiness in Adolescents. *Pediatrician*, 1990;17:5–12.

Carskadon, Mary, et al. Association Between Puberty and Delayed Phase Preference. *Sleep*, 1993;16(3):258–262.

Cartwright, Rosalind, and Lynne Lamberg. *Crisis Dreaming: Using Your Dreams to Solve Your Problems.* New York: HarperCollins, 1992.

Churchill, Winston. Quoted in Graebner, Walter, *My Dear Mister Churchill.* London: Michael Joseph, 1965, p. 55.

Colford, John M., Jr., and Stephen McPhee. The Ravelled Sleeve of Care: Managing the Stresses of Residency Training. *Journal of the American Medical Association*, 1989;261:889–893.

Dement, William, and Nathaniel Kleitman. The Relation of Eye Movements During Sleep to Dream Activity: An Objective Method for the Study of Dreaming. *Journal of Experimental Psychology*, 1957;43:339–346.

Dinges, David. The Benefits of a Nap During Prolonged Work and Wakefulness. *Work and Stress*, 1988;2(2):139–153.

———. Interview, October 1989.

———. When We Can and Cannot Judge Our Sleepiness on Awakening. *Sleep Research*, 1988;17:83.

Fiorini, William. Letters. *Journal of the American Medical Association*, 1988;259:2098.

Haslam, Diana. The Military Performance of Soldiers in Continuous Operations: Exercises "Early Call" I and II. In *Biological Rhythms, Sleep, and Shift Work, Advances in Sleep Research, Volume 7*, eds., Laverne C. Johnson, Donald I. Tepas, W. P. Colquhoun, and Michael J. Colligan. New York: SP Medical and Scientific Books, 1981, pp. 435–458.

Horne, James. Sleep Loss and Divergent Thinking Ability. *Sleep*, 1988;11:528–536.

Johnson, Laverne, and Paul Naitoh. The Operational Consequences of Sleep Deprivation and Sleep Deficit. (AGARD-AG-193), 1974.

Klass, Perri. *ZZZZZZ(Wha'?)ZZZZZZ(Huh?)ZZZZZZ*. *Discover*, December 1986:18–20.

Klebanoff, M.A., et al. Outcomes of Pregnancy in a National Sample of Resident Physicians. *New England Journal of Medicine*, 1990;323:1040–1045.

Kleitman, Nathaniel. Interview, June 1989.

Kupfer, David, ed. Sleep Disorders. In *Annual Review*, Vol. 4. Washington, D.C.: American Psychiatric Association, 1985, pp. 262–396.

Lavie, Peretz. To Nap, Perchance to Sleep—Ultradian Aspects of Napping. In *Sleep and Alertness: Chronobiological, Behavioral, and Medical Aspects of Napping*, eds., David Dinges and Roger Broughton. New York: Raven Press, 1989, pp. 99–120.

Lewthwaite, Gilbert. Clinton Spends First Day Playing Host. *The Baltimore Sun*, January 22, 1993:1A.

Lurie, Nicole, et al. How Do House Officers Spend Their Nights? *New England Journal of Medicine*, 1989;320:1673–1677.

Meddis, R., et al. An Extreme Case of Healthy Insomnia. *Electroencephalography and Clinical Neurophysiology*, 1973;35:213–214.

Mitler, Merrill, et al. Catastrophes, Sleep, and Public Policy: Consensus Report. *Sleep*, 1988;11:100–109.

Monk, Timothy. Circadian Aspects of Subjective Sleepiness: A Behavioural Messenger? In *Sleep, Sleepiness and Performance*, ed., Timothy Monk. Chichester, England: John Wiley and Sons, 1991, pp. 39–64.

National Center for Health Statistics. *Trends in Smoking, Alcohol Consumption, and Other Health Practices Among U.S. Adults: 1977 and 1983*. Hyattsville, Md.: U.S. Department of Health and Human Services, 1986.

National Commission on Sleep Disorders Research. *Wake-Up America: A National Sleep Alert*. Vol. 1: Executive Summary and Executive Report. Bethesda, Md.: National Institutes of Health, 1993, p. 48.

Rechtschaffen, Allan, et al. Sleep Deprivation in the Rat. *Sleep*, 1989;12(1):1–4.

Richardson, Gary. Interviews, June, July 1990.

Roehrs, Timothy. Interview, September 1989.

———, et al. Sleep Extension in Sleepy and Alert Normals. *Sleep*, 1989;12(5):449–457.

Rosekind, Mark, et al. Alertness Management in Flight Operations: Strategic Napping. Presentation at Aerospace Technology Conference and Exposition, Long Beach, California, September 23–26, 1991. Warrendale, Pa.: Society of Automotive Engineers, Inc., 1991.

———. Pilot Fatigue, Sleep, and Circadian Rhythms: NASA Fatigue Counter Measures Program. *Aviation Safety Journal*, 1993;3:20–24.

Shakespeare, William. *King Henry IV, Part II.* Act III, scene 1, line 5. *The Complete Works of Shakespeare*, ed., Hardin Craig. Chicago: Scott, Foresman and Company, 1961, p. 719.

Shapiro, C. M., and M. J. Flanigan. Function of Sleep. *British Medical Journal*, 1993:306(6874):383–385.

Sheehan, Neil. *A Bright Shining Lie: John Paul Vann and America in Vietnam.* New York: Random House, 1988.

Stampi, Claudio, and Bradford Davis. Forty-Eight Days on the "Leonardo da Vinci" Strategy for Sleep Reduction: Performance Behaviour with Three Hours Polyphasic Sleep Per Day. *Sleep Research*, 1991;20:471.

U.S. Congress, Office of Technology Assessment. *Biological Rhythms: Implications for the Worker.* OTA-BA-463. Washington, D.C.: U.S. Government Printing Office, September 1991, p. 178.

Webb, Wilse. *Sleep: The Gentle Tyrant.* Englewood Cliffs, N.J.: Prentice-Hall, Inc., 1975.

Webb, Wilse, and David Dinges. Cultural Perspectives on Napping and the Siesta. In *Sleep and Alertness: Chronobiological, Behavioral, and Medical Aspects of Napping*, eds., David Dinges and Roger Broughton. New York: Raven Press, 1989, p. 260.

Weiss, R. Safety Gets Short Shrift on Long Night Shift. *Science News*, 1989;135:37.

Wilson, A. Murray, and G. Weston. Application of Airline Pilots' Hours to Junior Doctors. *British Medical Journal*, 1989;299:779–781.

Wolff, George, and John Money. Relationship Between Sleep and Growth in Patients with Reversible Somatropin Deficiency (Psychosocial Dwarfism). *Psychological Medicine*, 1973;3:18–27.

Chapter 5: The Rhythms of Sexuality

American Psychiatric Association. Premenstrual Dysphoric Disorder Symptom Checklist. In *Diagnostic and Statistical Manual of Mental Disorders*, 4th Ed. Washington, D.C.: American Psychiatric Association, 1994.

Cutler, Winnifred Berg, et al. Human Axillary Secretions Influence Women's Menstrual Cycles: The Role of Donor Extract from Men. *Hormones and Behavior*, 1986;20:463–473.

Dewan. E. M. On the Possibility of a Perfect Rhythm Method of Birth Control by Periodic Light Stimulation. *American Journal of Obstetrics and Gynecology*, 1967;99:1016–1018.

Drennan, Michael. Night Lights Shorten Long and Irregular, But Not Regular, Menstrual Cycles. *Society for Light Therapy and Biologic Rhythms*, 1991;3:20.

Golub, Sharon. *Periods: From Menarche to Menopause.* Newbury Park, Calif.: Sage Publications, 1992, pp. 64–65, 69.

Karacan, Ismet, et al. The Ontogeny of Nocturnal Penile Tumescence. *Waking and Sleeping*, 1976;1:27–44.

Lee, Kathryn. Circadian Temperature Rhythms in Relation to Menstrual Cycle Phase. *Journal of Biological Rhythms*, 1988;3:255–263.

Lee, Kathryn, et al. Sleep Patterns Related to Menstrual Cycle Phase and Premenstrual Affective Symptoms. *Sleep*, 1990;13:403–409.

Levine, Richard, et al. Differences in the Quality of Semen in Outdoor Workers During Summer and Winter. *New England Journal of Medicine*, 1990;323:12–16.

Lin, May. Night Light Alters Menstrual Cycles. *Psychiatry Research*, 1990; 33:135–138.

McClintock, Martha. Menstrual Synchrony and Suppression. *Nature*, 1971;229:244–245.

Mills, Joyce. Premenstrual Syndrome: Symptom, or Source of Transformation? *Psychological Perspectives*, 1988;19:101–110.

Parry, Barbara. Morning Versus Evening Bright Light Treatment of Late Luteal Phase Dysphoric Disorder. *American Journal of Psychiatry*, 1989;146: 1215–1217.

Parry, Barbara, et al. Light Therapy of Late Luteal Phase Dysphoric Disorder: An Extended Study. *American Journal of Psychiatry*, 1993;150:1417–1419.

Premenstrual Syndrome. Washington, D.C.: American College of Obstetricians and Gynecologists. January 1989.

Preti, George, et al. Human Axillary Secretions Influence Women's Menstrual Cycles: The Role of Donor Extract of Females. *Hormones and Behavior*, 1986;20:474–482.

Reinberg, Alain, and Michel Lagoguey. Circadian and Circannual Rhythms in Sexual Activity and Plasma Hormones (FSH, LH, Testosterone) of Five Human Males. *Archives of Sexual Behavior*, 1978;7(1)13–30.

Rex, Katharine, et al. Light Treatment of Long, Irregular Menstrual Cycles. *Sleep Research*, 1993;22:414.

Roenneberg, Til, and Jürgen Aschoff. *Journal of Biological Rhythms*, 1990;5:217–239.

Severino, Sally, and Margaret Moline. *Premenstrual Syndrome.* New York: Guilford Press, 1989, p. vii.
Shaver, Joan, et al. Sleep Patterns and Stability in Perimenopausal Women. *Sleep,* 1988;11:556–561.

Chapter 6: Better Sleep

American Sleep Disorders Association. *The International Classification of Sleep Disorders: Diagnostic and Coding Manual.* Rochester, Minn.: American Sleep Disorders Association, 1990, pp. 117–140.
Ancoli-Israel, Sonia. Now I Lay Me Down to Sleep: The Problem of Sleep Fragmentation in Elderly and Demented Residents of Nursing Homes. *Bulletin of Clinical Neurosciences,* 1989;54:127–132.
Bliwise, Donald. Apparent Seasonal Variation in Sundowning Behavior in a Skilled Nursing Facility. *Sleep Research,* 1989;18:408.
Campbell, Scott, et al. Alleviation of Sleep Maintenance Insomnia with Timed Exposure to Bright Light. *Journal of the American Geriatric Society,* 1993;41:829–836.
Czeisler, Charles, et al. Bright Light Resets the Human Circadian Pacemaker Independent of the Timing of the Sleep-Wake Cycle. *Science,* 1986;233: 667–671.
———. Chronotherapy: Resetting the Circadian Clock of Patients with Delayed Sleep Phase Insomnia. *Sleep,* 1981;4:1–21.
deBeck, Thomas. Delayed Sleep Phase Syndrome—Criminal Offense in the Military? *Military Medicine,* 1990;155:14–15.
Editorial. When the Body Clock Goes Wrong: Delayed Sleep Phase Syndrome. *Lancet,* 1992;2:884.
Ferber, Richard. Interview, August 1988.
———. Sleep Schedule-Dependent Causes of Insomnia and Sleepiness in Middle Childhood and Adolescence. *Pediatrician,* 1990;17:13–30.
Hippocrates. *On Endemic Diseases (Air, Waters and Places),* Vol. 5. Eds., J. N. Mattock and M. C. Lyons. Cambridge, England: Heffer and Sons, 1969.
Klein, Torsten, et al. Circadian Sleep Regulation in the Absence of Light Perception: Chronic Non-24-hour Circadian Rhythm Sleep Disorder in a Blind Man with a Regular 24-Hour Sleep-Wake Schedule. *Sleep,* 1993:16(4):333–343.
Lingjaerde, O., et al. Insomnia During the "Dark Period" in Northern Norway. An Explorative, Controlled Trial with Light Treatment. *Acta Psychiatrica Scandinavica,* 1985;71:506–512.
Mant, Andrea, and E. Ann Eyland. Sleep Patterns and Problems in Elderly

General Practice Attenders: An Australian Survey. *Community Health Studies*, 1988;12:192–199.

Miles, Laughton, and William Dement. Sleep and Aging. *Sleep*, 1980;3: 119–220.

Miles, Laughton, et al. Blind Man Living in Normal Society Has Circadian Rhythms of 24.9 Hours. *Science*, 1977;198:421–423.

Monk, Timothy. Daily Social Rhythms in the Elderly and Their Relation to Objectively Recorded Sleep. *Sleep*, 1992;15:322–329.

Moore-Ede, Martin, et al. Circadian Timekeeping in Health and Disease. Part 2: Clinical Implications of Circadian Rhythmicity. *New England Journal of Medicine*, 1983;309:530–536.

National Sleep Foundation/Gallup Poll. *Sleep in America.* Los Angeles: National Sleep Foundation, 1991.

Pollak, Charles. Clinical and Social Consequences of Disordered Sleep. In *Abstracts of the NIH Consensus Development Conference on the Treatment of Sleep Disorders of Older People*, March 26–28, 1990. Bethesda, Md.: National Institutes of Health, 1990, pp. 107–109.

Pollak, Charles, and Deborah Perlick. Sleep Problems and Institutionalization of the Elderly. *Sleep Research*, 1987:16:407.

Reynolds, Charles, et al. Sleep and Chronobiologic Disturbances in Late Life. In *Geriatric Psychiatry*, eds., E. Busse and D. Blazer. Washington, D.C.: American Psychiatric Press, 1989, pp. 475–488.

Rosenthal, Norman, et al. Phase-Shifting Effects of Bright Morning Light as Treatment for Delayed Sleep Phase Syndrome. *Sleep*, 1990;13:354–361.

Sack, Robert, et al. Circadian Rhythm Abnormalities in Totally Blind People: Incidence and Clinical Significance. *Journal of Clinical Endocrinology and Metabolism*, 1992;75(1):127–134.

———. Melatonin Administration to Blind People: Phase Advances and Entrainment. *Journal of Biological Rhythms*, 1991;6(3):249–261.

Satlin, Andrew, et al. Bright Light Treatment of Behavioral and Sleep Disturbances in Patients with Alzheimer's Disease. *American Journal of Psychiatry*, 1992:149;1028–1032.

Seligmann, Jean, and Linda Buckley. A Sickroom with a View. *Newsweek*, March 26, 1990:61.

Spielman, Arthur, et al. Treatment of Insomnia by Restriction of Time in Bed. *Sleep*, 1987;10:45–56.

Stevenson, Jim. Interview, September 1989.

Terman, Michael, et al. Dawn and Dusk Simulation as a Therapeutic Intervention. *Biological Psychiatry*, 1989;25:966–970.

Chapter 7: The Rhythms of Sickness and Health

Barnes, Peter. Autonomic Control of the Airways and Nocturnal Asthma as a Basis for Drug Treatment. In *Chronopharmacology: Cellular and Biochemical Interactions*, ed., Björn Lemmer. New York: Marcel Dekker, Inc., 1989, pp. 53–63.

Bernard, C. *Leçons sur les phénomènes de la vie communs aux animaux et aux végétaux.* Paris: J. B. Bailliere, 1885.

Bolli, Geremia, and John Gerich. The Dawn Phenomenon—A Common Occurrence in Both Non-Insulin-Dependent and Insulin-Dependent Diabetes Mellitus. *New England Journal of Medicine*, 1984;310:746–750.

Burton, Robert. *The Anatomy of Melancholy.* Vol 1. New York: Dutton, 1961. (Originally published in 1621.)

Cavallini, Marco, et al. Chronobiology in the Service of Surgery. *Biochimica Clinica*, 1989;13:917–931.

Center for Chronic Disease Prevention and Health Promotion. Seasonality in Sudden Infant Death Syndrome—United States, 1980–1987. *Morbidity and Mortality Weekly Report*, 1990:39:891–895.

The Diabetes Control and Complications Trial Research Group. The Effect of Intensive Treatment of Diabetes on the Development and Progression of Long-term Complications in Insulin Dependent Diabetes Mellitus. *New England Journal of Medicine*, 1993:329:977–986.

Douglas, Neil. Asthma. In *Principles and Practice of Sleep Medicine*, eds., Meir H. Kryger, Thomas Roth, and William Dement. Philadelphia: W. B. Saunders Co., 1989, pp. 591–600.

Garland, Frank, Cedric Garland, et al. Geographic Variation in Breast Cancer Mortality in the U.S.; A Hypothesis Involving Exposure to Solar Radiation. *Preventive Medicine*, 1990;19:614–622.

Graeber, R. C., et al. *Human Eating Behavior: Preferences, Consumption Patterns, and Biorhythms.* Technical Report Natick/TR-78/022. Natick, Mass.: U.S. Army Natick Research and Development Command, 1978.

Halberg, Francine, et al. Chronobiology, Radiobiology and Steps toward the Timing of Cancer Radiotherapy. In *Cancer Management in Man: Detection, Diagnosis, Surgery, Radiology, Chronobiology, Endocrine Therapy*, ed., A. L. Goldson. Dordrecht, the Netherlands: Kluwer Academic Publishers, 1989, pp. 227–253.

Halberg, Franz. Interviews, January, July 1990.

———. Some Aspects of the Chronobiology of Nutrition: More Work Is Needed on "When to Eat." *Journal of Nutrition*, 1989;119:333–343.

Halberg, F., et al. Autorhythmometry—Procedures for Physiologic Self-Measurements and Their Analysis. *Physiology Teacher*, 1972;1:1–11.

———. Susceptibility Rhythm to E. *Coli* Endotoxin and Bioassay. *Proceedings of the Society for Experimental Biology and Medicine of New York*, 1960, 103:142–144.

———. Toward a Chronopsy: Part II. A Thermopsy Revealing Asymmetrical Circadian Variation in Surface Temperature of Human Female Breasts and Related Studies. *Chronobiologia*, 1979;6(3):231–257.

Hrushesky, William. Circadian Chronotherapy: From Animal Experiments to Human Cancer Chemotherapy. In *Chronopharmacology: Cellular and Biochemical Interactions*, ed., Björn Lemmer. New York: Marcel Dekker, Inc., 1989, pp. 439–473.

———. The Clinical Application of Chronobiology to Oncology. *American Journal of Anatomy*, 1983;168:519–542.

———. Menstrual Influence on Surgical Cure of Breast Cancer. *Lancet*, 1989:2:949–952.

Klevecz, R. R., and P. S. Braly. Circadian and Ultradian Rhythms of Proliferation in Human Ovarian Cancer. *Chronobiology International*, 1987;4(4): 513–523.

Kowanko, I. C., et al. Domiciliary Self-Measurement in Rheumatoid Arthritis and the Demonstration of Circadian Rhythmicity. *Annals of the Rheumatic Diseases*, 1982;41:453–455.

Laidlaw, J. Catamenial Epilepsy. *Lancet*, 1956;2:1235–1237.

Lamberg, Lynne. Chronotherapeutics: Implications for Drug Therapy. *American Pharmacy*, 1991;N831(11):20–23.

McCall, Nancy, et al. The Effect of Enteric-Coated Aspirin on the Morning Increase in Platelet Activity. *American Heart Journal*, 1991;121:1382–1388.

McCleary, Richard, et al. Age- and Sex-Specific Cycles in United States Suicides, 1973 to 1985. *American Journal of Public Health*, 1991;81:1494–1497.

Mahowald, Mark, et al. Sleep Fragmentation in Rheumatoid Arthritis. *Arthritis and Rheumatism*, 1989;32:974–983.

Manson, JoAnn, et al. A Prospective Study of Aspirin Use and Primary Prevention of Cardiovascular Disease in Women. *Journal of the American Medical Association*, 1991;266:521–527.

Marler, J. R., et al. Morning Increase in Onset of Ischemic Stroke. *Stroke*, 1989;20(4):473–476.

Martin, Richard. Nocturnal Asthma. In *Wake-up America: A National Sleep Alert*. Report of the National Commission on Sleep Disorders Research, Vol. 2. Bethesda, Md.: National Institutes of Health, 1994, pp. 99–105.

Mitler, Merrill, et al. Circadian Rhythm of Death Time: Cause of Death Versus Recorded Death Time in New York City. *Sleep Research*, 1985;14:306.

Moore-Ede, Martin. Physiology of the Circadian Timing System: Predictive Versus Reactive Homeostasis. *American Journal of Physiology*, 1986, May;250(5 Pt. 2):R737–752.

Muller, James, et al. Circadian Variation in the Frequency of Onset of Acute Myocardial Infarction. *New England Journal of Medicine*, 1985;313: 1315–1322.

Murdock, Barbara Scott. *Chronobiology*. Minneapolis: Earl Bakken, 1986.

National Center for Health Statistics. Deaths from Selected Causes, by Date of Death. United States, 1987. *General Mortality*, p. 315.

Office of Scientific and Health Reports, National Institute of Neurological and Communicative Disorders and Stroke. *Epilepsy: Hope Through Research*. Bethesda, Md.: National Institutes of Health, 1981.

Office of Scientific and Health Reports, National Institute of Neurological and Communicative Disorders and Stroke. *Headache: Hope Through Research*. Bethesda, Md.: National Institutes of Health, 1984.

Paccaud, Fred. Hour of Birth as a Prognostic Factor for Perinatal Death. *Lancet*, 1988;1:340–343.

Rabkin, Simon. Chronobiology of Cardiac Sudden Death in Man. *Journal of the American Medical Association*, 1980;244:1357–1358.

Reflections on the Rise in Asthma Morbidity and Mortality. Editorial. *Journal of the American Medical Association*, 1990;264:1718–1719.

Reinberg, A., and F. Levi. Clinical Chronopharmacology with Special Reference to NSAIDS. *Scandinavian Journal of Rheumatology*, 1989; (Suppl. 65):118–122.

Reinberg, Alain. Clinical Chronopharmacology: An Experimental Basis for Chronotherapy. In *Biological Rhythms and Medicine*, eds., Alain Reinberg and Michael Smolensky. New York: Springer-Verlag, 1983, pp. 209–263.

Reinberg, Alain, and Michael Smolensky. Introduction to Chronobiology. In *Biological Rhythms and Medicine*, eds., Alain Reinberg and Michael Smolensky. New York: Springer-Verlag, 1983, pp. 1–21.

Ridker, P. M, et al. Circadian Variation of Acute Myocardial Infarction and the Effect of Low-Dose Aspirin in a Randomized Trial of Physicians. *Circulation*, 1990;82:897–902.

Rivard, Georges, et al. Maintenance Chemotherapy for Childhood Acute Lymphoblastic Leukemia: Better in the Evening. *Lancet*, 1985;2: 1264–1266.

Shouse, Margaret. Epilepsy and Seizures During Sleep. In *Principles and Practice of Sleep Medicine*, eds., Meir H. Kryger, Thomas Roth, and William Dement. Philadelphia: W. B. Saunders Co., 1989, pp. 364–376.

Smolensky, Michael. Aspects of Human Chronopathology. In *Biological Rhythms and Medicine*, eds., Alain Reinberg and Michael Smolensky. New York: Springer-Verlag, 1983, pp. 131–209.

Smolensky, Michael, and Gilbert D'Alonzo. Medical Chronobiology: Concepts and Applications. *American Review of Respiratory Disease*, 1993;147:S2–S19.

Szabo, S. The Role of Gastric Secretion in the Chronopharmacology of Drug-induced Gastric and Duodenal Ulcers. Presentation at a meeting, *Clinical Applications of Chronobiology*, National Institutes of Health, Bethesda, Md.; June 20, 1989, pp. 17–18.

Weiss, Kevin. Seasonal Trends in US Asthma Hospitalizations and Mortality. *Journal of the American Medical Association*, 1990;263:2323–2328.

Chapter 8: Treating Depression

Cartwright, Rosalind, and Lynne Lamberg. *Crisis Dreaming: Using Your Dreams to Solve Your Problems.* New York: HarperCollins, 1992.

Cook, Frederick. Gynecology and Obstetrics Among the Eskimos. *Brooklyn Medical Journal*, 1894;8:154–169.

Eastman, Charmane. What the Placebo Literature Can Tell Us About Light Therapy for SAD. *Psychopharmacology Bulletin*, 1990;26:495–504.

Fitzgerald, F. Scott. *The Crack-up.* New York: New Directions, 1945.

Ford, Daniel, and Douglas Kamerow. *Journal of the American Medical Association*, 1989;262:1479–1484.

Gillin, J. Christian. The Sleep Therapies of Depression. *Progress in Neuropsychopharmacology and Biological Psychiatry*, 1983;7:351–364.

Goodwin, Frederick K., and Kay Redfield Jamison. *Manic Depressive Illness.* New York: Oxford University Press, 1990.

Hellekson, Carla. Interview, July 1989.

Jacobsen, Frederick. Seasonal Affective Disorder: A Review of the Syndrome and Its Public Health Implications. *American Journal of Public Health*, 1987;77:57–60.

Kasper, Siegfried. Phototherapy in Individuals with and Without Subsyndromal Seasonal Affective Disorder. *Archives of General Psychiatry*, 1989;46: 837–844.

Kern, Herbert. Interviews, June 1986, July 1989; personal communication, July 6, 1989.

Kern, Herbert, and Alfred Lewy. Corrections and Additions to the History of Light Therapy and Seasonal Affective Disorder. *Archives of General Psychiatry*, 1990;47:90–91.

Klerman, Gerald, and Myrna Weissman. Increasing Rates of Depression. *Journal of the American Medical Association*, 1989;261:2229–2235.

Kripke, D. F. Illumination Measurement and Phototherapy. *Sleep '88*, ed., J. Horne. Stuttgart, Germany: Gustav Fischer Verlag, 1989, pp. 248–249.

Kripke, Daniel, et al. Phototherapy for Nonseasonal Major Depressive Disorders. In *Seasonal Affective Disorders and Phototherapy*, eds., Norman E. Rosenthal and Mary C. Blehar. New York: The Guilford Press, 1989, pp. 342–356.

Kupfer, David. REM Latency: A Psychobiologic Marker for Primary Depressive Disease. *Biological Psychiatry*, 1976;11:159–174.

Kupfer, David, ed. Sleep Disorders. In *Annual Review*, Vol. 4. Washington, D.C.: American Psychiatric Association, 1985, pp. 262–396.

Lahmeyer, Henry, et al. Morning Light Treatment of SAD. *Society for Light Therapy and Biologic Rhythms*, 1990;2:17.

Lewy, Alfred, et al. Bright Artificial Light Treatment of a Manic-Depressive Patient with a Seasonal Mood Cycle. *American Journal of Psychiatry*, 1982;139:1496–1497.

National Center for Health Statistics. *Death Rates for 72 Selected Causes by Month: United States, 1979–87. Vital Statistics of the United States. Vol. II, Mortality*. Washington, D.C.: U.S. Government Printing Office, 1989.

National Center for Health Statistics. *Unpublished final data.* Hyattsville, Md.: The National Center for Health Statistics, 1990.

National Depressive and Manic-Depressive Association. *Bipolar Disorder: Self Portrait of an Illness.* Presentation at the American Psychiatric Association Annual Meeting, San Francisco, May 24, 1993.

Office of Medical Applications of Research. *Mood Disorders: Pharmacologic Prevention of Recurrences.* National Institutes of Health/National Institute of Mental Health Consensus Development Conference. Bethesda, Md.: National Institutes of Health, April 24–26, 1984.

Parry, Barbara. Light Therapy for Premenstrual Syndrome. *Society for Light Therapy and Biologic Rhythms.* Continuing Medical Education Course Abstract, 1993;5:14.

Parry, Barbara, and Thomas Wehr. Therapeutic Effect of Sleep Deprivation in Patients with Premenstrual Syndrome. *American Journal of Psychiatry*, 1987:144:808–810.

Price, Reynolds. *Clear Pictures.* New York: Atheneum, 1989, pp. 123–124.

Richter, P., et al. Light Imagination with Hypnotized Winter Depressed Patients. *Society for Light Treatment and Biologic Rhythms*, 1989;1:21.

Rosenthal, Leora, et al. Prevalence of Seasonal Affective Disorder at Four Latitudes. *Psychiatry Research*, 1990;31:131–144.

Rosenthal, Norman. Diagnosis and Treatment of Seasonal Affective Disorder. *Journal of the American Medical Association*, 1993:270:2717–2720.

Rosenthal, Norman, et al. Seasonal Affective Disorder: A Description of the Syndrome and Preliminary Findings with Light Therapy. *Archives of General Psychiatry*, 1984:41:72–80.

Rosenthal, Norman, et al. Seasonal Affective Disorder and Its Relevance for the Understanding and Treatment of Bulimia. In *Psychobiology of Bulimia*, eds. J. I. Hudson and H. G. Pope. Washington, D.C.: American Psychiatric Press, 1987, pp. 205–228.

Sack, David, et al. Potentiation of Antidepressant Medications by Phase Advance of the Sleep-Wake Cycle. *American Journal of Psychiatry*, 1985;142:606–608.

Siffre, Michael. Six Months Alone in a Cave. *National Geographic*, March 1975; 426–435.

Sobel, Dava. Interview, November 8, 1990.

———. Sleep Lab: subject No. 17 Heads Home. *The New York Times*, July 8, 1980:C1. (Series: June 17, June 24, July 1, July 8, July 15, 1980, all C1.)

Styron, William. *Darkness Visible: A Memoir of Madness*. New York: Random House, 1990.

———. Why Primo Levi Need Not Have Died. *The New York Times*, December 19, 1988, Section A, p. 17.

Tempest, Rone. The Longest Night. *Los Angeles Times*, December 9, 1988: Part V.

Terman, Michael, et al. Seasonal Symptom Patterns in New York: Patients and Population. In *Seasonal Affective Disorder*, eds., C. Thompson and T. Silverstone. London: CRC Clinical Neuroscience, 1989, pp. 77–97.

Ulrich, Roger. View Through a Window May Influence Recovery from Surgery. *Science*, 1984;224:420–421.

Vogel, Gerald, et al. Improvement of Depression by REM Sleep Deprivation. *Archives of General Psychiatry*, 1980;37:247–253.

Wehr, Thomas. Seasonal Affective Disorders: A Historical Overview. In *Seasonal Affective Disorders and Phototherapy*, eds., Norman E. Rosenthal and Mary C. Blehar. New York: The Guilford Press, 1989, pp. 2–32.

Wehr, Thomas, and David Sack. Sleep Disruption: A Treatment for Depression and a Cause of Mania. *Psychiatric Annals*, 1987;17:654–663.

Wehr, Thomas, et al. Eye Versus Skin Phototherapy of Seasonal Affective Disorder. *American Journal of Psychiatry*, 1987;144:753–757.

———. Phase Advance of the Circadian Sleep-Wake Cycle as an Antidepressant. *Science*, 1979;206:710–713.

———. Summer Depression: Description of the Syndrome and Comparison with Winter Depression. In *Seasonal Affective Disorders and Phototherapy*, eds., Norman E. Rosenthal and Mary C. Blehar. New York: The Guilford Press, 1989, pp. 55–63.

Woman in Isolation Test Is Found Dead. *Los Angeles Times*, January 19, 1990:A28.

Wirz-Justice, Anna. Light Therapy in Seasonal Affective Disorder Is Independent of Time of Day or Circadian Phase. *Archives of General Psychiatry*, 1993;50:929–937.

Wu, Joseph, and William Bunney. The Biological Basis of an Antidepressant Response to Sleep Deprivation and Relapse: Review and Hypothesis. *American Journal of Psychiatry*, 1990;1147:14–21.

Wurtman, Judith. Depression and Weight Gain: The Serotonin Connection. *Journal of Affective Disorders*, 1993;29:183–192.

Wurtman, Richard. The Effects of Light on Man and Other Mammals. *Annual Review of Physiology*, 1975;37:467–483.

Chapter 9: Coping with Jet Lag

Arendt, J., et al. Alleviation of Jet Lag by Melatonin: Preliminary Results of Controlled Double Blind Trial. *British Medical Journal*, May 3, 1986;292:1170.

Czeisler, Charles, and James Allan. Acute Circadian Phase Reversal in Man Via Bright Light Exposure; Application to Jet Lag. *Sleep Research*, 1987;16:605.

Czeisler, Charles, et al. Bright Light Induction of Strong (Type 0) Resetting of the Human Circadian Pacemaker. *Science*, 1989;244:1328–1333.

Ehret, Charles F., and Lynne Waller Scanlon. *Overcoming Jet Lag.* New York: Berkley Books, 1983.

Graeber, R. Curtis. The Impact of Transmeridian Flight on Deploying Soldiers. In *Biological Rhythms, Sleep, and Shift Work.* Advances in Sleep Research, Vol. 7, eds., Laverne Johnson, Donald Tepas, W. P. Colquhoun, and Michael Colligan. New York, SP Medical and Scientific Books, 1981.

———. Jet Lag and Sleep Disruption. In *Principles and Practice of Sleep Medicine*, eds., Meir H. Kryger, Thomas Roth, and William Dement. Philadelphia: W. B. Saunders Co., 1989, pp. 324–331.

Graeber, R. Curtis, ed. Sleep and Wakefulness in International Aircrews: A Cooperative Study. *Aviation, Space and Environmental Medicine*, 1986; 57(12,II):B1–B64.

In-Flight Survey of U.S. Travelers to Mexico and Overseas Countries. Washington, D.C.: U.S. Travel and Tourism Administration, August 1989.

Jahhar, P., et al. Psychiatric Morbidity and Time Zone Changes: A Study of Patients from Heathrow Airport. *British Journal of Psychiatry*, 1982:140:231–235.

Lavie, Peretz. The Effects of Midazolam 7.5 mg on the Resynchronization of the Temperature Cycle after Intercontinental Flights. *Sleep Research*, 1988;17:385.

Long, Michael E. What Is This Thing Called Sleep? *National Geographic*, 1987;12:787–821.

Moline, Margaret, et al. Effects of the "Jet Lag Diet" on the Adjustment to a Phase Advance. Abstract 111. *Society for Research on Biological Rhythms*, May 9–13, 1990.

Monk, Timothy, and Simon Folkard. Adjusting to the Changes to and from Daylight Saving Time. *Nature*, 1976;261:688–689.

Monk, Timothy, et al. Inducing Jet Lag in the Laboratory: Patterns of Adjustment to an Acute Shift in Routine. *Aviation, Space and Environmental Medicine*, 1988;56:703–710.

Neibuhr, Gustav. Toasting a Historic Papal Visit with Plenty of Ale Marys (Full of Taste). *Washington Post*, August 15, 1993:A18.

Post, W., and H. Gatty. *Around the World in Eight Days*. London: John Hamilton Ltd., 1931.

Servan-Schreiber, Jean-Louis. *The Art of Time*. Reading, Mass.: Addison-Wesley, 1988, p. 11.

Turek, Fred, and Susan Losee-Olsen. A Benzodiazepine Used in the Treatment of Insomnia Phase-Shifts the Mammalian Circadian Clock. *Nature*, 1986;321(6066):167–168.

Ward, Ritchie R. Von Frisch and Renner: The Clock of the Bees. In *The Living Clocks*. New York: New American Library, 1971, pp. 175–185.

Winget, Charles. A Review of Human Physiological and Performance Changes Associated with Desynchronosis of Biological Rhythms. *Aviation, Space and Environmental Medicine*, 1984;55:1085–1096.

Chapter 10: Work Around the Clock

Åkerstedt, Torbjörn. Sleepiness as a Consequence of Shift Work. *Sleep*, 1988;2:17–34.

Baker, Theodore, president, ShiftWork Systems, Inc. Interview, December 1993.

Bamford, James. The Last Flight of KAL 007. *Washington Post Magazine*, January 8, 1984:4–6, 8.

Biological Clocks and Shift Work Scheduling. Hearings Before the Subcommittee on Investigations and Oversight of the Committee on Science and Technology, House of Representatives, Ninety-eighth Congress, March 23–24, 1983. Washington, D.C.: U.S. Government Printing Office, 1983.

Carruthers, M., et al. Man in Transit: Biochemical and Physiological Changes During Intercontinental Flights. *Lancet*, 1976:1:977–980.

Cobb, Sidney, and Robert Rose. Hypertension, Peptic Ulcer, and Diabetes in Air Traffic Controllers. *Journal of the American Medical Association*, 1973;224:489–492.

Coleman, Richard. Interview, September 1987.

Czeisler, Charles. *Final Report on the Philadelphia Police Department Shift Rescheduling Program*. Boston: Center for Design of Industrial Schedules, January 28, 1988.

———. Personal communication, January 1992.

Czeisler, Charles, et al. Rotating Shift Work Schedules That Disrupt Sleep Are Improved by Applying Circadian Principles. *Science*, 1982;2117:460–463.

Dawson, Drew, and Scott Campbell. Timed Exposure to Bright Light Improves

Sleep and Alertness During Simulated Night Shifts. *Sleep*, 1991: 14:511–516.

Eastman, Charmane. Circadian Rhythms and Bright Light: Recommendations for Shift Work. *Work and Stress*, 1990;4:245–260.

———. High Intensity Light for Circadian Adaptation to a 12-H Shift of the Sleep Schedule. *American Journal of Physiology*, 1992;263:R428–R436.

Ehret, C. F., and D. R. Wernette. *Applying and Evaluating Chronobiological Findings in Shift-Work Industries.* 47th Ironmaking Conference Proceedings, Toronto Meeting, April 17–20, 1988. Warrendale, Pa.: Ironmaking Division of the Iron and Steel Society, Inc., 1988;47:191–201.

Flaim, Paul. Work Schedules of Americans: An Overview of New Findings. *Monthly Labor Review*, November 1986: 3–6.

Folkard, Simon, et al. Chronobiology and Shift Work: Current Issues and Trends. *Chronobiologia*, 1985;12:31–54.

———. Short and Long-Term Adjustment of Circadian Rhythms in "Permanent" Night Nurses. *Ergonomics*, 1978; 21:785–799.

Foushee, H. Clayton. Assessing Fatigue. *Airline Pilot*, May 1986:18–22.

French, J., J. P. Hannon, and G. C. Brainard. Effects of Bright Illuminance on Human Performance and Body Temperature. *Annual Review of Chronopharmacology*, 1990;7:37–40.

Gordon, Nancy, et al. The Prevalence and Health Impact of Shift Work. *American Journal of Public Health*, 1986;76:1225–1228.

Graeber, R. Curtis. Aircrew Fatigue and Circadian Rhythmicity. *Human Factors in Aviation*, eds., E. Wiener and D. C. Nagel. New York: Academic Press, 1987.

Hilliker, Nancy Ann Jenkins. Interview, October 15, 1992.

———. et al. Sleepiness/Alertness on a Simulated Night Shift Schedule and Morningness-Eveningness Tendency. *Sleep*, 1992;15:430–433.

Hochschild, Arlie. *The Second Shift, Working Parents and the Evolution at Home.* New York: Viking, 1989.

Klein, Marty, president, Synchrotech. Interview, February 1990.

Knutsson, Anders, et al. Increased Risk of Ischaemic Heart Disease in Shift Workers. *Lancet*, 1986;2:89–92.

Koen, Susan, president, MATRICES Consultants, Inc. Interview, February 1990.

Lauber, John, and Phyllis Kayten. Sleepiness, Circadian Dysrhythmia, and Fatigue in Transportation System Accidents. *Sleep*, 1988;11:503–512.

Lee, Kathryn. Interview, December 4, 1992.

———. Self-Reported Sleep Disturbances in Employed Women. *Sleep*, 1992;15:493–498.

———. Sleep-Related Health Problems Among Female Shiftworkers. *Sleep Research*, 1989;18:369.

Lesar, Timothy, et al. Medication Prescribing Errors in a Teaching Hospital. *Journal of the American Medical Association*, 1990;263:2329–2334.

Lewthwaite, Gilbert. Sailor in Charge of Ferry's Doors Admits Sleeping. The Baltimore Sun, May 2, 1987.

Melbin, Murray. *Night as Frontier: Colonizing the World After Dark.* New York: The Free Press, 1987.

Mitler, Merrill, et al. Catastrophes, Sleep, and Public Policy: Consensus Report. *Sleep*, 1988;11:100–109.

Monk, Timothy, and Donald Tepas. Shift Work. In *Job Stress and Blue Collar Work*, eds., C. L. Cooper and M. J. Smith. Chichester, England: John Wiley and Sons Ltd., 1985, pp. 65–84.

Monk, Timothy, and Simon Folkard. *Making Shift Work Tolerable*. London: Taylor and Francis, 1992.

Moore-Ede, Martin. *The Economic Impact of Human Maladaptation to Round-the-Clock Operations.* Boston: Institute for Circadian Physiology, 1988.

———. Jet Lag, Shift Work, and Maladaption. *News in Physiological Sciences*, 1986;1:156–157.

Moore-Ede, Martin, et al. Mathematical Simulation of the Effects of Rotating Shift Work Schedules on Circadian Sleep-Wake Cycles and Alertness Rhythms. *Sleep Research*, 1986;15:279.

National Academy of Sciences. Sleep, Biological Clocks, and Health. In *Health and Behavior: A Report of the Institute of Medicine*. Washington, D.C.: National Academy Press, 1982, pp. 109–116.

National Transportation Safety Board. *Aircraft Accident Report: China Airlines Boeing 747-SP, N4522V, 300 Nautical Miles Northwest of San Francisco, California, February 19, 1985. NTSB/AAR-86/03.* Washington, D.C.: National Transportation Safety Board, 1985.

National Transportation Safety Board. *Fatigue, Alcohol, Other Drugs, and Medical Factors in Fatal-to-the Driver Heavy Truck Crashes. Vol. I. PB90-917002. NTSB/SS-90/01.* Washington, D.C.: National Transportation Safety Board, 1990, p. 87.

Nurminen, T. Shift Work, Fetal Development and Course of Pregnancy. *Scandinavian Journal of Work and Environmental Health*, 1989;15:395–403.

Orth-Gomér, Kristina. Intervention on Coronary Risk Factors by Adapting a Shift Work Schedule to Biologic Rhythmicity. *Psychosomatic Medicine*, 1983;45(5):407–415.

Powers, Leslie, et al. Bright Light Treatment of Night-Shift Workers. *Society for Light Treatment and Biological Rhythms*, 1989;1:34.

Presser, Harriet, and Virginia Cain. Shift Work Among Dual-Earner Couples with Children. *Science*, 1983;219;876–879.

Richardson, Gary. Interviews, September 1987, April 1993.

Rosa, Roger, et al. Extended Workdays: Effects of 8-Hour and 12-Hour Rotating

Shift Schedules on Performance, Subjective Alertness, Sleep Patterns and Psychosocial Variables. *Work and Stress*, 1989;3:21–32.

Rutenfranz, Joseph, et al. Occupational Health Measures for Nightworkers and Shift Workers. In *Hours of Work*, eds., Simon Folkard and Timothy Monk. Chichester, England: John Wiley and Sons Ltd., 1985, pp. 199–210.

Sayle, Murray. Closing the File on Flight 007. *The New Yorker*, December 13, 1993:90–101.

Scherrer, J. Man's Work and Circadian Rhythm through the Ages. In *Night and Shift Work: Biological and Social Aspects*, eds., Alain Reinberg, et al. Oxford, England: Pergamon Press, 1981, pp. 1–10.

Siwolop, Sana, et al. Helping Workers Stay Awake at the Switch. *Business Week*, December 8, 1986:108.

Stam, James, executive vice-president, Circadian Technologies, Inc. Interview, September 1989.

Stewart, K. T., et al. Light Treatment for NASA Shift Workers. Fifth International Conference of Chronopharmacology and Chronotherapeutics. Amelia Island, Florida, July 12–16, 1992:II-9.

Tepas, Donald. Flexitime, Compressed Workweeks and Other Alternative Work Schedules. In *Hours of Work*, eds., Simon Folkard and Timothy Monk. Chichester, England: John Wiley and Sons Ltd., 1985, pp. 147–164.

Tepas, Donald, et al. Changing Industry to Continuous Operations: Different Strokes for Different Plants. *Behavior Research Methods, Instruments, and Computers*, 1985;17(6):670–676.

———. Shiftwork and the Older Worker. *Experimental Aging Research*, 1993;19:295–320.

U.S. Congress, Office of Technology Assessment. *Biological Rhythms: Implications for the Worker*. OTA-BA-463. Washington, D.C.: U.S. Government Printing Office, September 1991, pp. 76, 95, 106, 155, 207.

U.S. Congress, Office of Technology Assessment. *Gearing Up for Safety: Motor Carrier Safety in a Competitive Environment*, OTA-SET-382. Washington, D.C.: U.S. Government Printing Office, September 1988.

U.S. Congress, Office of Technology Assessment. *Safe Skies for Tomorrow: Aviation Safety in a Competitive Environment*, OTA-SET-381. Washington, D.C.: U.S. Government Printing Office, July 1988.

Webb, Wilse. Are There Permanent Effects of Night Shift Work on Sleep? *Biological Psychology*, 1983;16:273–283.

Zurawik, David. Interview, March 1990.

Chapter 11: A Call for Public Policy Change

Campbell, James. Personal communication. September 23, 1989.

Council of Teaching Hospitals Housestaff Survey. Washington, D.C.: Association of American Medical Colleges, 1989, pp. 26, 40.

European Communities. Council Directive 93/104/EC Concerning Certain Aspects of the Organization of Working Time. *Official Journal of the European Communities,* Brussels, Belgium, November 23, 1993, pp. L307/18-L307/24.

Matsumoto, Kazuya. Effects of Nighttime Nap and Age on Sleep Patterns of Shift Workers. *Sleep,* 1987;10(6):580–589.

Monk, Timothy. *How to Make Shiftwork Safe and Productive.* Des Plaines, Ill.: American Society of Safety Engineers, 1988.

National Commission on Sleep Disorders Research, *Wake-Up America: A National Sleep Alert.* Vol. 1: Executive Summary and Executive Report. Bethesda, Md.: National Institutes of Health, 1993, pp. 23, 58.

Stevens, Amy. Bosses Fret They May Be Liable for Tired Workers on Road Home. *The Wall Street Journal,* April 16, 1991:B1, B9.

U.S. Congress, Office of Technology Assessment. *Biological Rhythms: Implications for the Worker.* OTA-BA-463. Washington, D.C.: U.S. Government Printing Office, September 1991, pp. 123–140, 168, 199, 200, 208.

Appendix: Strategies for Self-Help

Hauri, Peter, and Shirley Linde. *No More Sleepless Nights.* New York: John Wiley and Sons, 1990.

Monk, Timothy. *How to Make Shiftwork Safe and Productive.* Des Plaines, Ill.: American Society of Safety Engineers, 1988.

U.S. Congress, Office of Technology Assessment. *Biological Rhythms: Implications for the Worker.* OTA-BA-463. Washington, D.C.: U.S. Government Printing Office, September 1991, p. 107.

SHIFT WORK CONSULTING FIRMS

Circadian Technologies, Inc.
One Alewife Center
Cambridge, MA 02140-2317
(617) 492-5060
FAX: (617) 492-1442
Contact: Martin Moore-Ede, M.D., Ph.D.

Coleman Consulting Group, Inc.
P.O. Box 128
32 Ross Common
Ross, CA 94957-0128
(415) 461-5555
FAX: (415) 461-5591
Contact: Richard Coleman, Ph.D.

General Chronobionics, Inc.
410 S. Madison St.
Hinsdale, IL 60521
(708) 323-5642
FAX: (708) 323-5642
Contact: Charles Ehret, Ph.D.

MATRICES Consultants, Inc.
136 Commercial St.
Portland, ME 04101-4744
(207) 828-5151
FAX: (207) 828-5156
Contact: Susan Koen, Ph.D.

ShiftWork Consultants, Inc.
P.O. Box 604673
Bayside, NY 11360-4673
(718) 224-2767
FAX: (718) 229-5590
Contact: Jack Connolly, Ph.D.

ShiftWork Systems, Inc.
One Kendall Square
Building 200, 4th Floor
Cambridge, MA 02139
(617) 374-9340
FAX: (617) 374-9210
Contact: Theodore Baker, Ph.D.

SynchroTech
315 S. 9th St., Suite 211
Lincoln, NE 68508
(402) 474-4387
FAX: (402) 474-4425
Contact: Marty Klein, Ph.D.

Work Systems Research
P.O. Box 426
Mansfield Center, CT 06250
(203) 429-6260
FAX: (203) 486-2760
Contact: Donald Tepas, Ph.D.

INDEX